Chemical and Physical Methods for Protecting Biopolymers against Pests

CHEMICAL AND PHYSICAL METHODS FOR PROTECTING BIOPOLYMERS AGAINST PESTS

O. S. Kukovinets, M. I. Abdullin, R. A. Zainullin, R. V. Kunakova and G. E. Zaikov

Nova Biomedical Books
New York

For permission to use material from this book please contact us:
Telephone 631-231-7269; Fax 631-231-8175
Web Site: http://www.novapublishers.com

Library of Congress Cataloging-in-Publication Data

Chemical and physical methods for protecting biopolymers against pests / O.S. Kukovinets ... [et al.].

p. cm.

ISBN 978-1-60456-331-3 (hardcover)

1. Biopolymers--Deterioration. 2. Pesticides. 3. Pests--Control. I. Kukovinets, O.S.

QP801.B69C54 2008

572'.33--dc22

2008000039

Published by Nova Science Publishers, Inc. ✦ New York

Contents

Preface

Protection of natural biopolymers: proteins (collagen, keratin), carbohydrates (starch, cellulose) and others, is a very acute problem. Numerous fungi, insects and rodents damage raw materials, and finished products both during manufacturing and at storage. To study methods and protection facilities we systematized information on main types of biodestructors and gave numerous examples of synthesis of the most known and active preparations to control insects and other organisms damaging natural polymers in light and food industries.

The monograph is written by a group of authors famous for their papers on chemistry of low-molecular bioregulators and means for plant protection. It is designed for specialists engaged in problems of biopolymer protection.

Introduction

Pesticides (from Latin *pestis* – pest + *caedere* – kill) are chemical compounds applied for protection of plants, agricultural products, wood, articles made of wool, cotton, leather, for killing animal ectoparasites and for combating carriers of serious diseases.

Agricultural activities of a human being have resulted in concentration of large amounts of proteins and carbohydrates in comparatively small area. Due to this fact there appeared favorable conditions for development of different kinds of pests then forced to spend a lot of time and efforts to find means of subsistence. Since ancient times, manufacturers and consumers of various biopolymer products and materials have been searching for ways and means of protection thereof against pests.

With the development of chemistry it became possible to make efficient preparations widely applied in combating pests and rodents. New categories of insecticide-active organic compounds were found, e.g. organophosphorous pesticides and analogs of natural substances: juvenile hormones, insect pheromones, pyrethroids, etc. Table 1 shows that in 1990 among 150 applied insecticides 6 were inorganic, but in 2000 there were 3 such insecticides left only. Over 50% of all applied categories of compounds account for organophosphorous chemical pest-killers and pyrethroids. It should be mentioned therewith the decrease of total amount of preparations from 150 to 104, which can testify to creation of more active and effective preparations, as well as to development of more efficient ways of applications thereof making it possible to achieve better results through less number of preparations.

Fumigants (methyl bromide, methallyl chloride, dichloroethane, phostoxin, phosphine) and contact chemical pest-killers (carbophos, trichlormetaphos-3, metathion, volaton, actellic, malathion) are often used for combating grain insects and grain and cereal disinfection. Practically all these preparations possess acute toxicity for both man and warm-blooded animals, besides there is a risk of chronic poisoning of animals and people using treated grain as food: up to 1.0 mg/kg of carbophos is permitted in groats and bread [1]. Methyl bromide contained in the composition of widely advertised mixture of methyl bromide and phosphine is referred to as ozone-depleting compounds [2, 3]. Regular application of known preparations facilitates insecticide resistance with insects. Such circumstances in the course of time will undoubtedly result in full ban for applying highly toxic and environment-damaging pesticides in countries with developed agriculture and will promote for scientific research in finding new forms and kinds of highly active and environmentally safe preparations.

Table 1. Dynamics of spectrum of insecticides and acaricides (1990-2000)

Chemical group of preparations	1990		2000	
	Number of preparations	reactant	Number of preparations	reactant
Inorganic	6	5	3	1
Chloroorganic	8	6	-	-
Bromic	2	2	1	1
Nitro-containing	3	3	2	2
Carbamates	13	8	8	5
Organophosphorous compounds	59	24	27	10
Pyrethroids	41	16	31	12
Nereistoxin	1	1	1	1
Neonicotinoids	1	1	5	4
Phenylpyrazoles	-	-	4	1
Mixtures	3	2	5	5
Other groups	13	12	17	12

*Acaricide – a chemical preparation for plant protection against harmful mites.

This monograph contains information on up-to-date methods for protecting agricultural products (biopolymers of vegetable and animal origin) at storing and processing, methods for preparation synthesis and ways for successful application in practice.

Chapter 1

1. Living Organisms Damaging Biopolymer Materials

Among general reasons for damadging stored tissues, leather, furs and other biopolymers the first is high humidity, sharp temperature fluctuations. However these factors could not be so harmful if they did not provoke to putrefaction, mold appearance i.e. consequences of microorganisms activity (microscopic mold fungi). Besides, great harm can be caused by more highly organized things, mainly Arthropoda of insects and chiggers class. One must know basics of biology of harmful organisms to efficiently find methods for protecting fur and leather-processing production.

Base unit of all vegetable and animal organisms is a cell (Figure 1). Elementary organisms contain one cell, when their structure is complicated cells start making various essential functions and thus forming tissues and bodies. Cell contains cytoplasm where basic cellular organelle and chromosomes are found, i.e. genetic mechanism providing procreation, or cells either scattered without any limitations or contained in a nucleus.

By their systematic position these organisms belong to eukaryotes (nuclear) unlike prokaryotes (prenuclear) represented by bacteria and cyanobacteriae. It is availability of a nucleus, strictly limited by a membrane from cell contents, containing heredity basis i.e. a set of chromosomes, defines difference between eukaryotes and prokaryotes. Chromosomes of *y* prokaryotes are within nucleoplasm. They do not separate from the remaining part of a living cell. Prokaryotes have become the first living organisms appeared on Earth, while eukaryotes are the result of saltation which revealed infinite prospects for development before the world.

Despite the fact that prokaryotes, due to simplicity of structure and reproduction, and ability to use different substrates as energy source which are even of inorganic origin, are widely distributed, they are considered as cladotype (relict) in their current position.

Nucleation has become the beginning for differentiation deepening and making conditions for appearance of multicellular organisms whose development and complication in the course of evolution has resulted in diversification of the current living world.

Basis for differences between eukaryotes kingdom is way of how they use energy resources, i.e. nutrition. Animals are heterotrophs, i.e. they get substances necessary for their development from the outside only together with nutrition. But plants, especially higher ones, belong to autotrophs as they synthesize all organic substances necessary for life activity by themselves through photoactive pigments (chlorophyll, carotene) using sunlight as the energy source, and inorganic components (water, carbon dioxide, inorganic compounds of nitrogen, phosphorus, potassium and other elements) as building materials. However, there are many heterotrophs among plants leading either saprophytic (i.e. upon organic but not living substrates), or parasitic (upon living organisms) mode of existence.

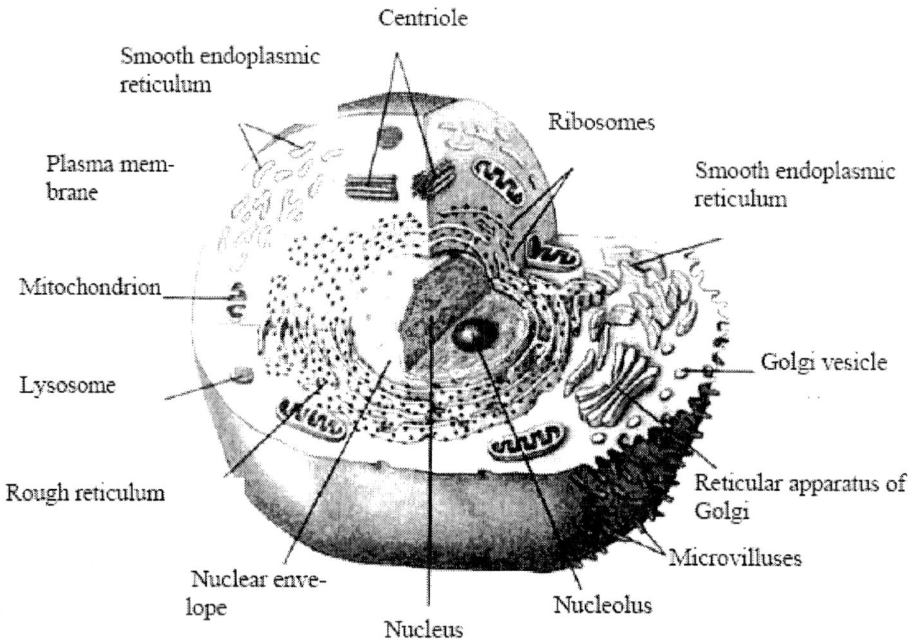

Figure 1. Structure of living organism cell.

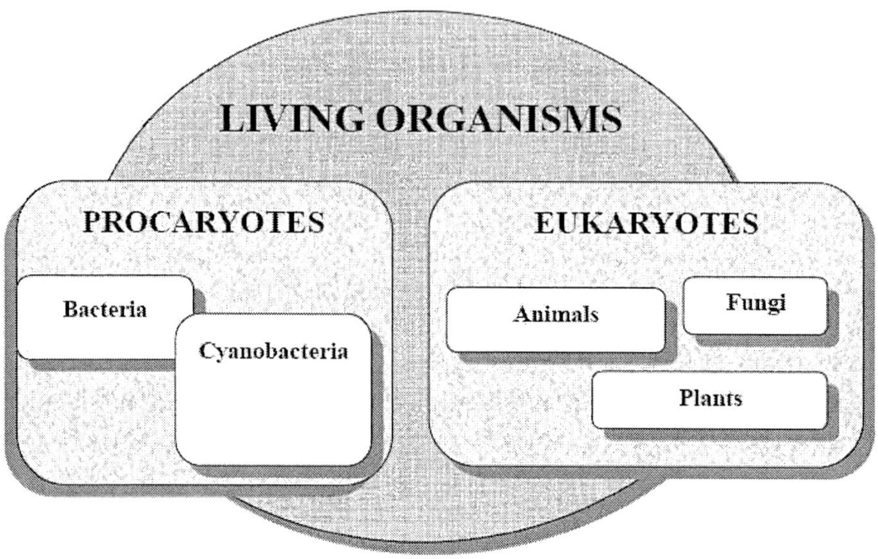

Figure 2. Systematics of living world.

Principal distinctions between animals and plants are availability of rigid cytoderm in plants and ability to active movements and change of places in animals. There are also some metabolic processes. Scientific systematic considers fungi as a separate kingdom of living organisms (Figure 2), although still recently they were included to plant kingdom within sub-kingdom of inferior plants.

1.1. Unicellular Organisms

Among protozoa one can also find pests of biopolymers of grain and cereals. Bacteria of *B.Subtilis ssp. mesentericus* (potato bacillus) subspecies being developed in bread-crumb cause potato disease. Potato bacillus occurs widely in nature. It is found in soil, air, plants, and others. These bacteria cause main problems in flour mills and bakeries.

If grain is not washed before grinding but moistened during grinding, potato bacillus gets into flour. Under favorable conditions potato bacillus bacteria are promptly reproduced. Temperature of about 40°C, moisture, nutrient medium, and sub-acidity are optimum conditions for development of potato bacillus spores. Its cells die when heated up to 80°C, but its spores are still viable at 120°C. Thus, bacteria die at baking, but their spores remain viable.

Acid medium inhibits development of potato bacillus bacteria. Therefore, potato disease rarely occurs in rye-bread with acidity.

Failure to comply with sanitary and technological rules for storage and treatment of grain, flour, as well as baking of bread and its storage significantly affects potato bacillus reproduction. Due to this much attention is paid to observance of current sanitary and technological rules applicable in grain-elevators, flour-milling and baking industries, as well as in trade.

With development of potato disease, reproduction of potato bacillus bacteria is strongly intensified. Due to affect of active amylases of potato bacillus, amount of dextrins* making a bread-crumb stick, is increased. Protein breakdown products formed therewith under influence of proteolytic enzymes of potato bacillus possess sharp specific smell. Potato disease-affected bread has sticky crumb which is stretched by threads if strongly affected and then in the middle of the loaf there is a black cavity with strong putrefactive odor.

To prevent spread of potato disease certain measures in all steps (soil, grain, bread) are to be conducted. For prevention thereof wheat flour ground down at mills is to be checked for potato bacillus through sample laboratory baking. The resulting molded bread in 1.5-2 hours after baking is wrapped in moistened porous paper in two layers and is placed in thermostat for 24 hours at 36-38°C and relative moisture of 83-87%. In 24 hours afterwards bread is unwrapped, cut up and checked for disease indications (specific smell, sticky crumb), then it is inspected in 36 hours.

Flour affected by potato disease is to be immediately realized, and all transportation links, equipment and premises of the mill and bakery are to be sanitized and disinfected. Surfaces are to be cleaned by 3% solution of acetic acid. Besides, all doors, windows, panels,

* Dextrines are oligomerhomologs formed under partial hydrolysis of linear regular homopolysaccharides.

floors and windows are cleaned by soap-wetted rags, then 3% solution of chlorinated lime and hot water ablution is applied.

1.2. Microscopic Fungi

Availability of cytoderm within fungi cells, as well as sap vacuoles contained within cells; protoplasm current well visible under microscope; absence of ability to active movements make it all possible to consider them close to plants, although fungi are heterotrophs (i.e. hemoorganoheterotrophs using chemically processed organic substances) by nutrition way. Fungi have no chlorophyll and its cell organelles – plastids).

Occurrence. Just few kinds of fungi can be found in aqueous medium, while they are widely spread overland. Saprophytes are actually found everywhere, where there can be organic substrates available as food. Generic diversification of saprophytic fungi is enormous. It includes mold fungi known elsewhere and edible fungi also known to everyone.

Parasitic species are widely numerous. They are capable to live using either plants or animals, including a man, as their hosts. Many of them are serious disease-producing organisms. Parasitic fungi give rise to outbreak of cultivated plants disease (epiphytotics) resulting in significant crop losses as well as to mycosis with animals and man difficult to cure. Pathogens of powdery mildew, black grain and ringworm are examples of this species.

General texture. Fungi are weakly differentiated multicellular living organisms with a simple texture. Fungus body consists of fibers (hyphae) and is called thallus. All range of thallus fibers is called mycelium (Figure 3).

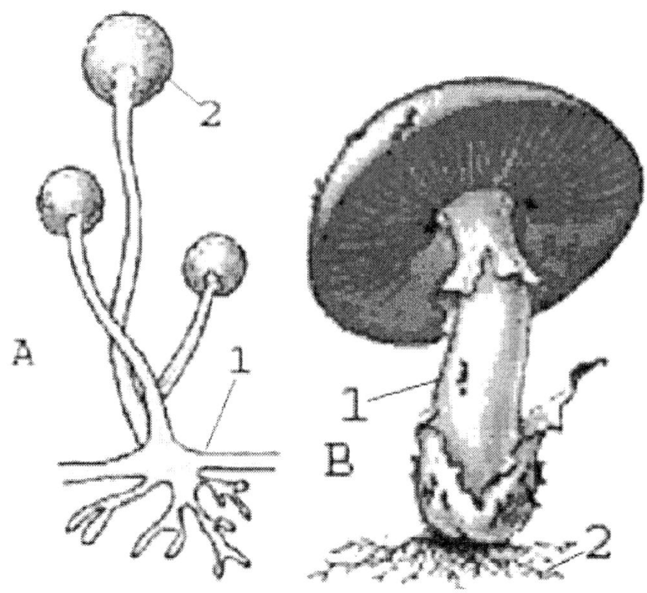

Figure 3. Fungi structure. A. Lower fungi: 1. Mycelium-formed hyphae. 2. Mycothallus. B. Higher fungi: 1. Mycothallus. 2.Rhizomorphs.

At some steps of fungi mycelium development there appear dense interlacements resembling tissue called plectenchyma (false parenchyma which is regarded as real tissue made of living cells). For example, edible mycothallus of blewits consists of plectenchyma. Differentiation of functions of parts of an organism in higher fungi resulted in appearance of rhizoid formations i.e. rhizomorphs also mycelium-formed.

Fungi hyphae possess cytoderms, and one can see vacuoles with cellular fluids and nucleus within through a microscope. Higher fungi and some kinds of lower ones contain transverse partitions dividing cytoplasm i.e. septa. Cytoplasm of neighboring cells is connected through pores contained within septa centers. Septa are not available with the majority of lower fungi.

Fungi cytoderm is dense and clearly expressed and usually contains chitin (N-acetyl-D-glucosamine polymer), and rarely cellulose.

Vegetative phase (period when there is growth only without cloning) with lower fungi is characterized by absence of cytoderm and formation of plasmodium i.e. multinuclear mobile protoplasmatic mass surrounded by thin membrane only, or pseudoplasmodium i.e. aggregation of mononuclear cells also possessing membrane only. There is no plasmodium or pseudoplasmodium in higher fungi, and vegetative phase is represented by hyphae only.

Systematic. Before being allocated to a separate kingdom fungi were included into sub-kingdom of lower plants as a division. Fungi classification is based upon both their phylogenetic (evolutionary-kindred) relations and practical aims. Fungi nomenclature is binary, it includes name of genus and species; e.g. mold fungus mucor known to everyone has generic name *Mucor* and specific name *mucedo*. Similar species are grouped into genera, genera into families *(–aceae*), families into orders (*–ales*), orders into classes (*-mycetes*). Mold fungi of practical importance for fabrics, leather and furs manufacture are grouped to zygomycetes class (lower fungi) and ascomycetes (higher fungi) (Table 2).

Table 2. Main fungi groups

Lower fungi (Phycomycetes)	Higher fungi (Eukomycetes)
Chytridiomycetes	Ascomycetes (cup fungi)
Oomycetes	Basidiomycetes
Zygomycetes	Deuteromycetes (imperfect fungi)

Reproduction. Fungi are reproduced both agamogenetically and syngenetically, with both of these methods and peculiarities thereof being the most important basis for fungi classification.

Fungi hyphae grow apically, with any part of mycelium being able to grow independently thus forming new thallome.

Fungi agamogenesis is possible by simple cell division into two parts, like bacteria, but there are such methods as pullulation (pinching with further full separation of mother cell, or outgrowth generated through preformed convexity into which one of nuclei passes after regular mitosis[*]) typical for yeast; hypha decay into separate cells; and finally spore formation.

When reproduced through hyphae decay into pieces these separate cells are called oidia or arthrospores. If these spores are surrounded by thick-walled shell they are called chlamydospores. If these spores are placed in special capsules i.e. sporangia, they are called sporangiospores.

When spores are formed through pitching of the upper part of hypha they are called conidiospores. Hyphae bearing sporangia or conidia are called sporangiophores and conidiospores accordingly. Spores of lower fungi keeping to water mode of life are active, have filaments and are called zoospores.

Fungi gamogenesis includes mandatory nuclear fusion of parental germ cells i.e. gametes containing half (haploid) number of chromosomes. When gametes are invisible they are called isogametes. Gametes of some fungi are formed in special differentiated cells i.e. gametangia. If gametangia are differentiated by form, then male gametangia are called antheridia and female ones are oogonia. If both male and female gametes are formed on one thallus from one original cell, then fungi with this way of syngenesis are considered homothallic, while if there are sex differences between thalluses they are called heterothallic.

Fungi are differentiated by way of gamete joining: lower fungi have both active gametes (planogametes), and their joining (conjugation) takes place outside gametangia.

Oomycetes have active male gamete entering oogonia and fertilizing ovules, and conjugation of polykaric gametangia is seen in zygomycetes.

Higher fungi have two forms of syngenesis. The first one is through fertilizing of female reproductive structure by antheridium, with no nuclei fusion taking place initially thus forming divisible dikaryons from which after the fusion of nuclei during fertilization (karyogamy) asci with spores (ascomycetes) are formed. The second one is through joining two cells of vegetative mycelium i.e. somatogamy with formation of basidiospores (basidiomycetes). Syngenesis of imperfect fungi is replaced by parasexual process where nuclear fusion takes place after their transition to another cell.

Genera *Aspergillus* and *Penicillum* are grouped into cup fungi. Cup fungi (ascomycetes) are divided into three groups depending on carposome forms (ascocarpus):

- Plectomycetes having absolutely closed carposomes (cleistothecia);
- Pyrenomycetes having flask carposomes (perithecia);
- Discomycetes – open, cotyloid carposomes (apothecia).

Aspergillus and *Penicillum* are grouped into plectomycetes. These genera are differentiated in conidia formation; if *Penicillum* conidia are formed directly on vertically growing conidiophore, then *Aspergillus* conidia are found on sterigmas to be formed on follicles, which vertically standing hyphae are ended with (Figure 4).

Microscopic fungi as biodecomposers. When textiles, leather and furs are stored in conditions of high humidity, mold being colonies of microscopic fungi can appear thereon.

[*] Mitosis (from Greek mitosis – thread) is indirect cell and cell body division. Due to mitosis a number of cells is increased.

More often they grow and develop on fabrics made of natural fibers and genuine leather, however, there are some species capable to live in synthetic fabrics and leatherette.

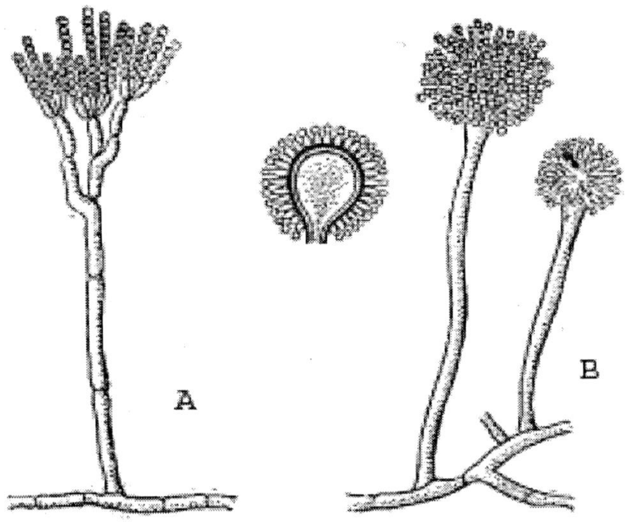

Figure 4. Forms of fruiting hypha with *Penicillum* (A) and *Aspergillus* (B).

Molds alone are not able to spoil textiles, leather and furs: destruction thereof is caused by fungi metabolic products i.e. metabolites liberated to environment, more exactly at substrate. It is essential that fungi metabolites are actively liberated when there is deficiency in any substance important for their activity. Organic acids and enzymes, liberated thereby, formed by fungi make the most destructive (deleterious) influence. Fungi make gluconic, lactic, oxalic, fumaric, succinic, malic, formic, acetic, and citric acids more often. The latter ones destroy natural organic materials and may cause destruction of synthetic fibers, changing structure and color thereof. Acids are formed involving other group of substances i.e. enzymes, which are very important for destruction. Main enzymes (exoenzymes) liberated by fungi belong to several groups:

- Oxidoreductases (oxygenases, dehydrogenases and oxydases): glucose oxydase, catalyze, peroxidase. Oxidoreductases catalyze redox reactions in the process of energy metabolism. Catalyze and peroxidase catalyze oxidation of different organic compounds by peroxide, mainly by hydrogen peroxide.
- Hydrolases – many of them are exoenzymes, they make substrata ready to be affected by other enzyme groups. They catalyze breakdown of ester bonds (esterases), bonds between glucose residues (glycosidases), peptide (amide) bonds (proteinases) by hydrolysis type.
- Lyases catalyze non-hydrolytic splitting of organic substances.
- Transferring enzymes catalyze transfer of atomic groups or parts of molecules from one sort of compounds to other ones.
- Isomerizing enzymes catalyze intramolecular transfer of radicals, groups, atoms thus forming isomers.

- Ligases (synthetases) catalyze synthesis of complex compounds from more simple ones.

Quiescent stages of microscopic mold fungi – spores – are easily scattered by air current and can survive for a long time in water, soil and other substrata not suitable for development. Thick-walled shell of spores makes them tolerant to both low and high temperatures, and ultraviolet radiation. Current chemical means for destroying fungi are fungicides which can be of either biological origin but mainly they are synthetic substances.

Usual methods of thermal heating – pasteurization (heating up to 60°C) and sterilization applying wet steam – boiling, autoclaving, or dry heat – warming up to 160-200°C cannot be applied for protecting textile, tanning, and fur industries goods. Thus, it is chemical means for protecting stored goods that are very important.

To choose the most efficient preparation one must define what kind of mold fungus does affect the stored materials. The attached table (table 3) presents development trends, the most typical substrates and description of fungi colonies. Combination of these features will help to define species of microscopic fungus-decomposer.

Table 3. Fungi affecting materials and tissues at storage

Species, genus	Substrate	Colony type	Development trends
Red-brown bissochlamys	Cotton yarn, paper, chamois leather, feather	From straw to brown-snuff color	High temperature (30-37°C)
Verrucate amauroascus	Putrefied leather	White color with further darkening	High temperature and humidity
Eurotium repens	Textiles, cellophane, plastics, rubber	Friable, red-orange with dark ring close to the edge	Lower humidity (13-15%), can grow at high salt content
Talaromyces flavus	Different fabrics and materials	Friable, with broken edge, of straw color with concentric orange rings	Can grow at high temperature
Chaetomium (different species)	Paper, cotton, ready-made clothes	Virescent, dark-primrose	High temperature and humidity
Penicillium purpurogenum	Paper, cotton fabrics, calico, leatherette	Velvet, dark-green with yellow border, the opposite side being purple-red	High humidity; can be found in soil, and grow at low temperature.
Penicillium expansum	Different materials	Velvet, with concentric yellow and green rings	
Aspergillus niger	Paper, leather, cotton fabrics	Brown or black color	High humidity
Aspergillus fumigatus	Wool, cotton	Velvetweed, cambridge blue or velvet, bluish-green	Affected materials are strongly heated when damaged by fungus
Aspergillus versicolor	Leatherette, calico, leather, canvas	Dense, raised, dark-green or bluish, often with pink border, the opposite side being rich red or cherry	Antiseptic-resistant
Fusarium	Calico, leatherette, leather	White, white-pink, pink-lilac or brown	High humidity and heat

1.3. Insects

Among living organisms found on the Earth, the most numerous class is insects. Insects are represented by the majority species harmful to man in all spheres of activity, including production and storage of biopolymers.

1.3.1. Taxonomy

Insect class is subdivided into two subclasses:

- apterygots;
- pterygots.

Subclasses are subdivided into orders and further to families. Apterygots subclass includes 4 insect orders, while pterygots subclass contains over 25 orders (Figure 5).

Insect species capable to damage tissues, leather, and finished goods at storage belong to lepidoptera orders (moths), coleopterous (dermestids), cockroach, hymenoptera (ants), and to termites – in southern regions.

- Insects are the most important object of scientific experiments (e.g. drosophila in genetics, confused flour beetle in population genetics, etc.).
- Flour beetle *Tribolium castaneum*is a widely-spread grain pest, has become the next organism which genome is being deciphered by American scientists. Researchers, who have received millions of dollars, use free time of genome deciphering machines which were engaged previously for studying human genome*. It is explained by the beetle practicality in studying main biological processes and in elaboration of new methods for pest control. All genetic information on the beetle is supposed to be placed in Internet after completion of the work. The deciphering process is headed by Suzan Brown from Kansas University. According to preliminary information flour beetle genome consists of 200 million base pairs, containing in 11 chromosomes. This DNA encodes thousands of genes.

1.3.2. Body Structure

Insect body which is better seen in larvae consists of 12-15 separate transverse segments connected in such a way that insect body (Figure 6) is separated into three parts: head, thorax and abdomen. Region separating thorax and abdomen is not often seen with adult insects and it cannot be seen in many larvae (e.g. that of flies). The peculiar feature of insect class is that adult species have three pairs of legs only. It is mentioned in one of class names: *Hexapoda* – Hexapod.

* Genome (from Greek *genos* – generation, origin): set of genes contained in a single set of cell chromosome.

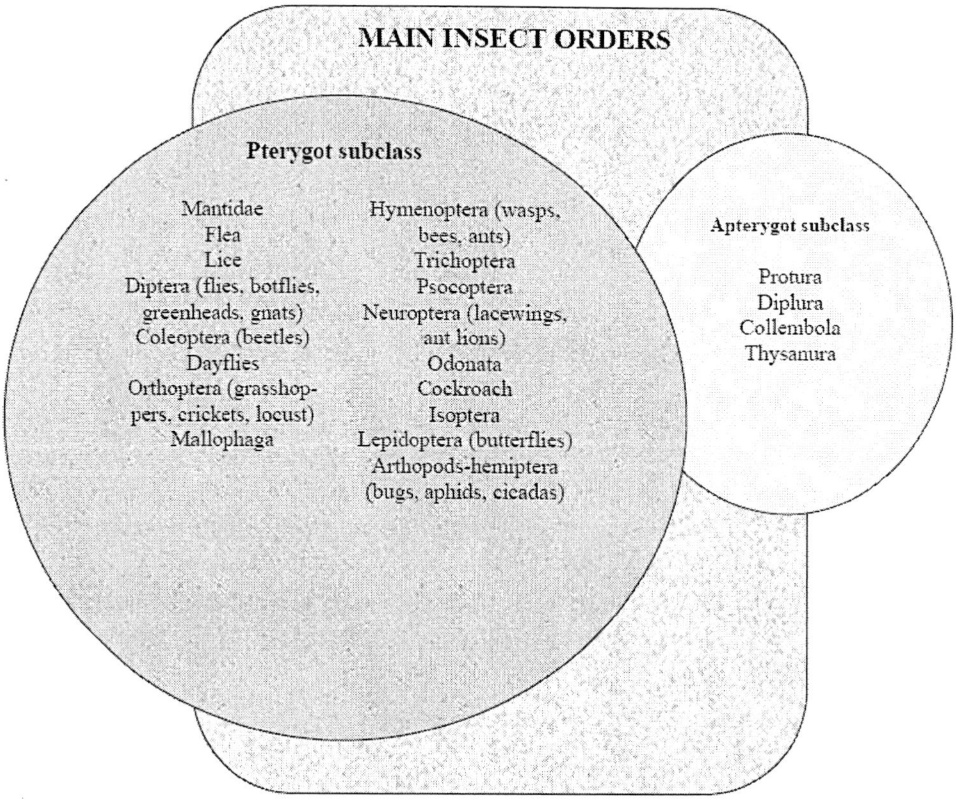

Figure 5. Main insect orders.

Adult insect head has a pair of compound eyes and three simple eyes (ocelli). Some insect larvae are blind (flies, mosquitoes). Visual acuity of insects is lower than that of humans, however, they can distinguish movements and forms of objects as well as some colors. As a rule, insect mouthparts have complex structure and variety thereof within one class is great, being adapted to mode of existence and pabular substrates of each species and even each insect instar. The following main patterns of mouthparts are recognized:

Biting pattern. Primarily typical for species consuming solid food. This pattern is present in most of agricultural pests feeding on plants (phytophages), as well as pests of industrial production under storage – dermestids, moth larvae, cockroaches and ants. Biting mouthpart pattern is characterized by strong jaws (mandibles) with sharp edges capable to active food shredding. Similar pattern is present in some predatory insects.

Rasping pattern. Found in species feeding by not so solid substrates; it consists of well developed glossa, often compound one and covered by small arms capable to rasp off food substrate particles. This rasping pattern is seen first of all in flies.

Sucking pattern. Merging sections of mouthpart form proboscis often equal to body length (of a butterfly), when extended. Insects with such mouthpart feed on nectar collected from flowers.

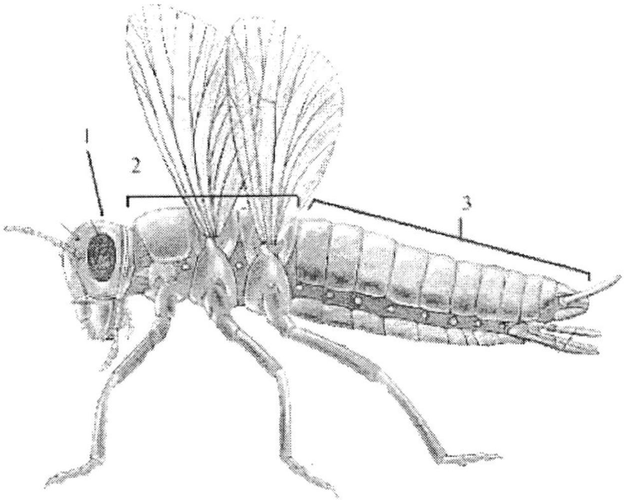

Figure 6. General body structure. 1.Head. 2.Thorax. 3.Abdomen.

Quite often mouthparts have special features making them very complex. In addition to the above there are other variations such as rasping-licking structures (bees) and biting-sucking ones with strong and sharp proboscis that is formed of several components, with part thereof being more hard and sharp (stylets), piercing hard and elastic surfaces (hemipterans, aphids, mosquitoes).

Very often different insect instars were developed in different environment and fed with different food; accordingly their mouthparts are different. Such differences are clearly demonstrated by representatives of Lepidoptera, hymenoptera and diptera orders. Among Lepidoptera there are species which adults do not feed at all and their mouthparts are not developed. For example, moths belong to such species, including clothes moth, being one of main pests of stored production. Their larvae, which mouthparts make it possible to utilize practically any fibers, are more harmful.

Apart from eyes and mouthparts insect heads have antennae i.e. binate sense organs which are often referred to everywhere as feelers. These feelers contain smell sensors of hypersensitivity. Due to them insects are successful in food searching as well as in perceiving smells over long distances produced by opposite individuals (sex pheromones), as well as other smells of vital importance (alarm pheromones produced by one species individuals; aggregation pheromones allowing one species individuals to gather together which is often necessary for normal evolution and reproduction; kairomones i.e. smells attracting predators and pests to victims).

Insect taste sensors, located not only on parts of the mouthpart but on tarsus surface, are also very sensible. Insects have organs of hearing, however, they are well developed with some species only which adults are capable to make loud enough sounds (e.g. crickets, grasshoppers). Hearing organs are located either on insect tarsi or abdomen.

Insect thorax is primarily intended for making active movements: lower part thereof possesses three pairs of legs, and the upper one – wings formed by outer covering folds. Insect thorax includes bulk of muscles responsible for movements of legs and wings. Insect legs are well developed and consist of several portions with flexible joints between (haunch, trochanter, hip, shank, and tarsus).

Insect tarsi terminate in claws and other pad-like structures helping them to keep to even vertical surface and to freely move over it (e.g. pneumatic suculi in fly tarsi). Depending on mode of existence legs can perform different functions (Figure 7). Mainly it is running, however, pair of hind-legs can become natatorial with applanate shank covered with long bristles densely located along the edge, as in water-tigers or in some other water insects. In grasshoppers, locusts and other Orthoptera a pair of hind-legs is hopping – haunch consists of strong muscles, shank is long, a joint between them provides springy strong push.

Pair of fore-legs can be used for digging paths: in such insects as mole crickets fore-legs are distinguished by well developed bucket-like tarsus, shank is also strong with comb-shaped hard outgrowths, thus helping mole crickets to make paths under the ground. Fore-legs can assist predators to seize and hold captures (e.g. Mantidae), with the fore-legs possessing hard and sharp acanthoid outgrowths.

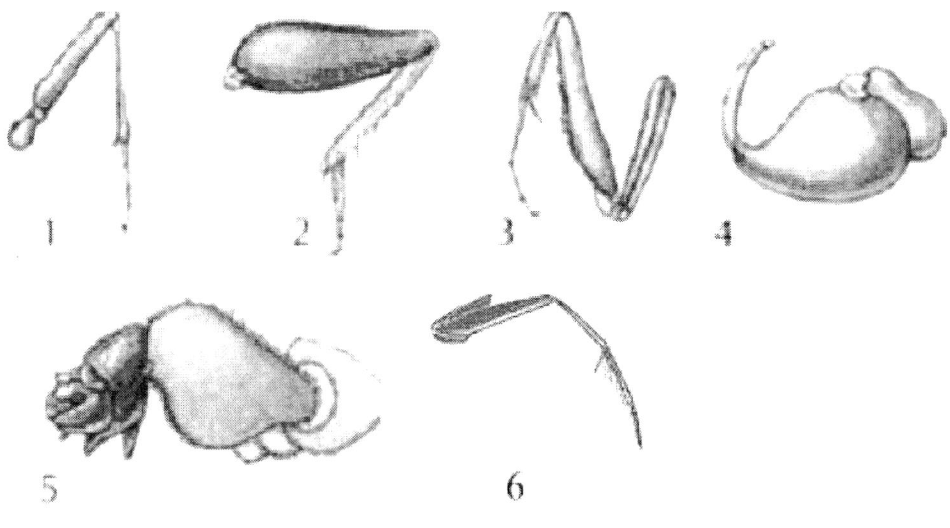

Figure 7. Types of insect legs. 1. Running or ambulatory leg. 2. Hopping leg. 3, 4. Predatory leg. 5. Digging leg. 6. Swimming leg.

Insect wings basically are intended for locomotion. There are typically two pairs of wings), but in a few groups (Diptera having one pair of well-developed wings) they are reduced to a single pair. As a rule, hindwings are thin, transparent or semitransparent, soft membrane with a numerous longitudinal and cross veins. These veins contain nerve cords and respiratory vessels (trachea) connected with appropriate systems of insect body. Wings are operated due to contraction of thorax flight muscles and thus they can make great number of complex movements. In representatives of Coleoptera order a pair of forewings is modified to form shell-like protactive elytra, which can rise and put sideways but they do not take part in active wing movements as they are hard enough and have small surface area. Elytra in some Hemiptera, Orthoptera, cockroaches and mantids are similar in function but rigid only partially. In parasitic bugs (e.g. cimex) both pairs of wings are fully reduced.

Insect abdomen consists of segments, lower part thereof can contain pseudopods, which availability makes it possible to differentiate between caterpillars (caterpillar larvae) and bug larvae, sometimes very similar externally. At the end of the abdomen in some insects female there is an ovipositor down which eggs pass to a substrate at the time of egg laying (e.g. Orthoptera oviposit to soil). Some species that are characterized by open eggs disposition female ovipositor is not developed (butterflies, beetles).

1.2.3. The Most Important Functional Systems of Insects

The whole insect body including extremities is protected by *covers* being dense coating consisting of internal layer – hypoderma formed by epithelium cells, and external one – cuticle having complex composition based upon excretions from hypoderma cells; protein basis of cuticle is connected with chitin providing rigidity to covers. Insects have no bony skeleton which serves as locomotor apparatus in vertebrate, however, their stiff and strong covers replace bony skeleton and therefore they are regarded as exoskeleton, i.e. skeleton inside the body. Covers of insect larvae do not possess such rigidity as the adult ones and can have soft membranes attaching neighboring parts of the body. All that makes it possible for larvae to greatly increase body size during development.

Insect covers are often of different coloration. Coloration can be either protective similar to surface color where insects spend majority of their life; sematic coloration, as a rule, it is very bright and contrasting considered by predators as inedible. For species where visual orientation is of great importance coloration can be alluring (e.g. light wings of female gipsy moths are regarded as visual orientation by male ones).

Insect *perigastrium* is made by covers and muscles attached thereto. Muscles of insects like Arthropoda order differ from muscles of all other invertebrates: they are barred thus providing far more rate of contractions as compared with unstriated musculature. Relative power of insect muscles is quite great: a man can lift a load of up to 0.8 of man's body, and a bee can carry mass 23 times exceeding its weight.

Insect perigastrium contains vital organs that can be identified as circulatory, nervous, digestive, excretory, reproductive and hormonal (endocrine) systems.

Insect *circulatory system* is open. It means that insect blood (haemolymph) comes freely through all internal bodies. However, haemolymph is particularly directed specified by heart – i.e. the most important body of the circulatory system. Insects possess only one large blood vessel – dorsal, and the heart is its enlargement located mainly in the back part of the vessel. Heart is divided into a number of chambers with ostia through which haemolymph enters heart from body cavity (Figure 8).

Heartbeatings are caused by muscular work. These muscles are located segment-wise by chambers of heart, and due to their work contractions of the blood vessel then force the haemolymph forward along its length to discharge in the head region.

Direction of haemolymph within the body cavity is provided by contractile partitions (diaphragms); with the whole system of valves of dorsal vessel participating therein.

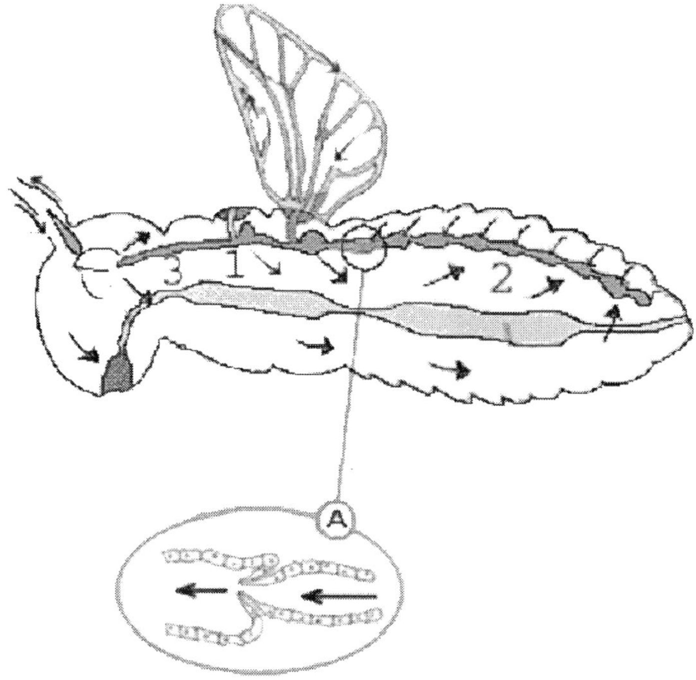

Figure 8. Insect circulatory system. 1.Dorsal vessel. 2.Chambers of heart. 3.Aorta. A – enlarged view of heart with valves. Arrows indicate direction of haemolymph flow within the body cavity.

Initially haemolymph enters head cavity passing through brain located in it, and then enters middle and hind body segments. Haemolymph acts like human blood distributing food materials from alimentary organs, it circulates hormones from vascular glands to other organs, and pigments contained in blood cells (hemocytes) like human hemoglobin (hemocyanin) participate in carrying oxygen, providing respiration. Haemolymph also serves for carrying away metabolic products from insects.

Haemolymph cells are also the most important part of insect protective system being involved in any defense reactions either passively (providing release of different secretions to haemolymph), or actively – ingesting foreign cells (phagocytosis).

Adipose body cells are functionally linked with haemolymph. This lacinated organ basically consists of adipose cells (adipocytes), synthesis of many essential proteins (transport proteins possessing hormonal, or trophic or protective properties) takes place in adipose body. Adipose body operates like depositing organ making the same functions as human liver.

Insect *respiratory system* is a system of tracheae i.e. branched air tubes coming though all bodies (Figure 9). Tracheae provide free air access to many sections of the insect body therefore their respiratory function of haemolymph (transfer of oxygen and carbon dioxide) is lower than that of vertebrates.

Air enters tracheae of terrestrial insects through spiracles (stigmata) i.e. paired apertures with closing valves located along insect abdomen and thorax. Some water species, as well as larvae and pupae leading water way of existence have spiracles closed, but thin membrane closing them provides bilateral gas diffusion both in and out of the body cavity. Many water insects have gills i.e. cuticular outgrowths of spiracle membranes penetrated by finest tracheae (tracheoles).

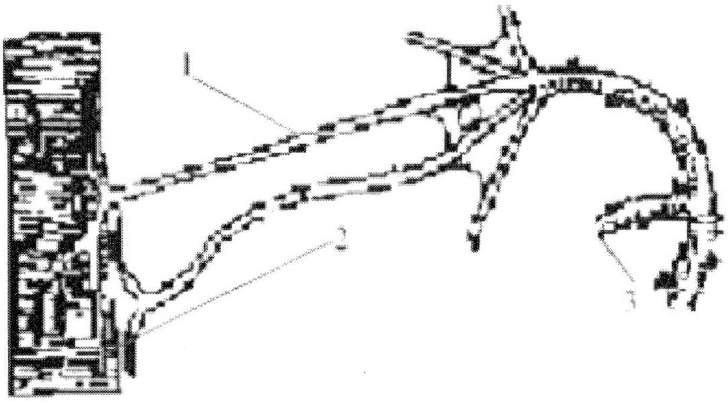

Figure 9. Insect respiratory system (scheme). 1.Cross tracheal trunks. 2.Lengthway tracheal trunks. 3.Spiracles.

Muscle contraction of abdomen of terrestrial insects assist to active air exchange. In flying insects tracheae form large expansions with thin walls to be filled with air at inhalation and decreased at exhalation.

Digestive system is presented by alimentary canal running from the mouth to the anus (Figure 10). Alimentary canal is divided into foregut, midgut and hindgut different by structure and functions. Foregut serves as oesophagus, it possesses chitinized lining. Midgut visually differs from other sections of intestine: it is much wider and possesses rather marked musculature. Developed epithelium covering midgut region, secretes great number of digestive enzymes. Besides it is involved in formation of peritrophic membrane – delicate but rather strong membrane covering midgut inner surface. Peritrophic membrane protects organism against pathogenic bacteria, and acts like one-sided filter for digestive enzymes concentrated around the food to be digested; it is present in all species except those with extraintestinal digestion.

Hindgut is divided in its turn into small gut and rectum. Process of intake of food to be digested is completed in small gut cavity. Main function of rectum on the whole is absorption of water from food and formation of dehydrated excrements. Like foregut the rectum also possesses chitinized lining and is intended for dejection to be excreted from intestine cavity. Intestine of insects feed on vegetable food hard to be digested is longer than that of predators or parasites intestine. Some species (e.g. termites) possess simbionts within intestine i.e. bacteria which feed cellulose. Insect *excretory system* is presented by blindly finished outgrowths located in the area between mid- and hind intestine i.e. malpighian tubules. Malpighian tubules are intended for filtration of moisture from midgut cavity. But it is rectum which performs excretory function.

Insect *nervous system* is well developed (Figure 11) and presented by nerve network consisting of double nerve cords (ganglia) connected by longitudinal and cross bridges.

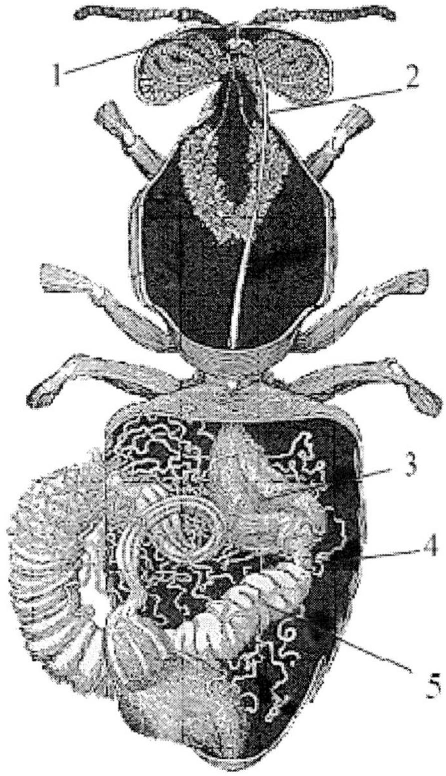

Figure 10. Digestive system of insects. 1.Mouthpart. 2.Foregut. 3.Midgut. 4.Small gut. 5.Rectum.

Central nervous system is divided into sections by body segments: supraesophageal ganglion - 1 in the head segment, thoracic ganglia – 2, and ventral nervous system -3.

Supraesophageal ganglion is the largest among all insect ganlia. It surrounds oesophagus by a ring and has a complex structure, and in the most developed groups of species it includes fore-part, mid-part and hind-part and their outgoing ophthalmic and antennal nerves. Functionally supraesophageal ganglion of insects is similar to vertebrate brain.

Thoracic ganglia are fused together; nerves outgoing therefrom innervate wings and legs. Ventral nervous system consists of geminated ganglia located segment-wise. Their nerve endings innervate all organs in the insect abdomen.

Developed brain and sense organs of insects make them adapt to different environmental conditions and changes thereof by realizing both chain of complex hereditary instincts and behavior reflexes appearing in the course of lifecycle.

Reproductive system of insects is intended for their mass reproduction. Both male and female have geminated viscera located in the abdominal cavity – testes and ovaries, respectively, where gametes are matured; their paired excretory ducts are combined into unpaired one before the last segment of the abdomen, which is expanded in female making generative cell, highly evolved species have spermatheca. It is in this generative cell, at least in terrestrial species, eggs are fertilized. Both male and female possess accessory germ glands. Their secretions contribute to either preservation of sperm viability, or cocoon formation protecting oviposition.

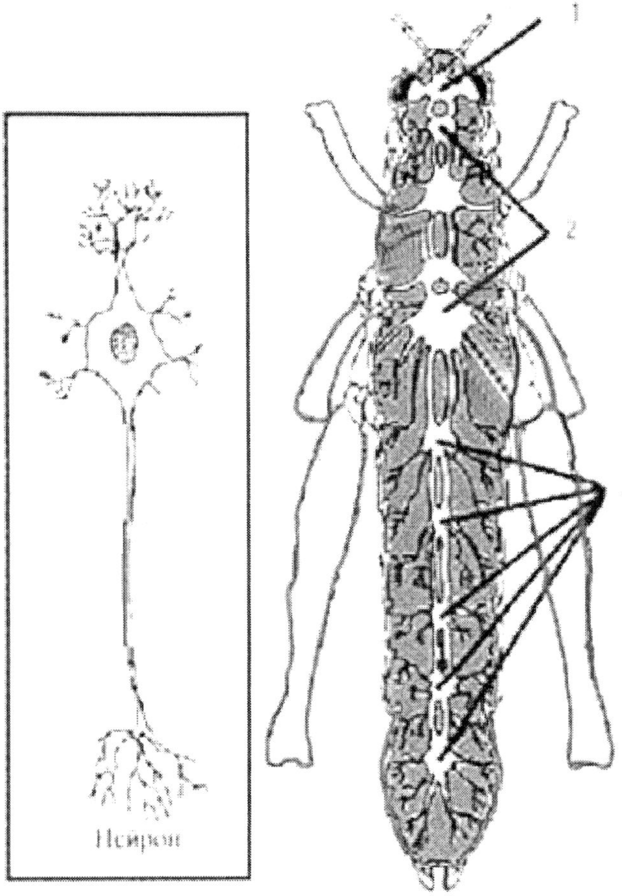

Figure 11. Nervous system of insects: 1.Supraesophageal ganglion. 2. Thoracic ganglia. 3. Ventral nervous system.

Endocrine (hormonal) system of insects is represented by a number of endocrine glands (Figure 12), to which belong adjacent and cardiac bodies forming retrocerebral complex adjoining rear part of brain, as well as prothoracic glands located in thorax.

Separate cells of middle part of brain – i.e. neurosecretory cells (NSC) function like endocrine organs. Wide set of hormones produced by insect endocrine glands regulates the whole metabolism, as well as provides control over evolution and vital functions. Neurosecretory cells excrete a number of hormones of peptide structure, with the most important of them being activation hormone generally stimulating activity of all other endocrine organs. Juvenile hormones (JH) are synthesized in adjacent bodies, which stimulate larvae development. Several juvenile hormones are found in insects, their balance being defined by a number of larval ages.

High content of juvenile hormones results in appearance of additional larval ages and does not allow larval stage development to be successfully completed: even if pupal molt (or transformation to the adult form) takes place, there appears species with a large number of larval features after it; such an insect will die soon.

JH are important for reproduction processes, they regulate ovule maturation in females and the beginning of active reproduction periods.

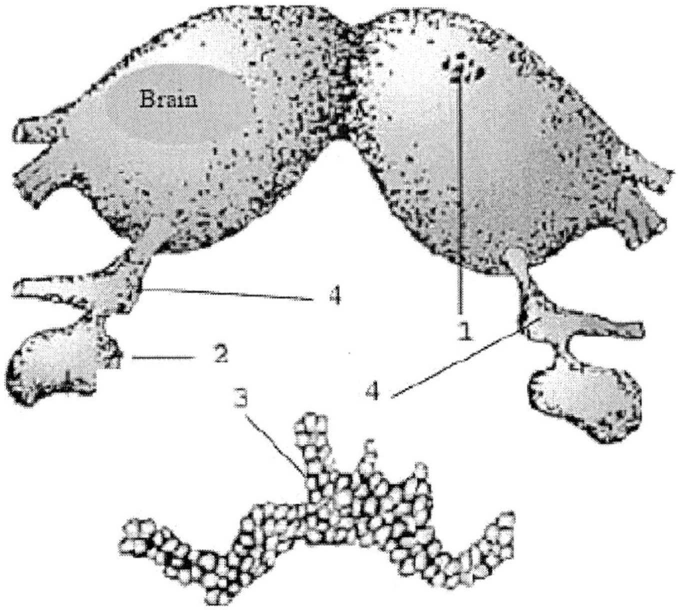

Figure 12. Endocrine system of insects. 1. Brain neurosecretory cells (NSC). 2.Adjacent bodies. 3. Part of prothoracic gland. 4. Cardiac bodies.

Major hormone synthesized in prothoracic gland – is of steroid nature. Insect ecdysones provide for beginning and normal passing of larval molting, so they are often called as larval hormones. Besides, these hormones regulate metabolism, as well as they can directly influence genetic mechanism, i.e. chromosomes.

Completion of insect development is regulated by pupal hormone – bursicon, as well as by imago hormone. A number of hormones is necessary for finer controlling metabolism process. For example, hyperglycemic hormone regulating carbohydrate metabolism, and adipokinetic hormoneregulating fat metabolism.

In social insects (e.g. bees) hormonal balance is a key factor determining which functions are to be done by adult insect in the family, therewith not only internal (endogenous) hormones are important but also the ones which come to the insect organism from outside, like royal jelly of bees, called exohormones and ectohormones. Such substances are usually synthesized by special glands (mandibular glands in bees).

Insect *development* includes three main steps: ovification (embryonal), metamorphosis and adult stage development (imago).

Embryonal stage takes place in ovum protected against outside influences by rigid capsule and sometimes by cocoon additionally. By duration this stage is strongly different in different insects: so, it takes typhoid fly egg about 6-8 hours to pass through this stage, and gipsy moth larvae (caterpillars) appear only next spring from eggs laid by females. During embryonal development there are periods of both active development and suspended activity i.e. diapauses.

In some species there can be seen viviparity where eggs are developed within the oviduct of females. Such kind of embryonic development is more typical for parasitic flies.

Larvae appearing from eggs very actively feed on and shortly increase both in weight and in body size; it is connected with availability of molt in insects. New, more elastic covers of the body are formed at moulting, after this old covers are flaked and shed as molting skin. Chitinized lining of oesophagus, hind intestine and trachea are also interchanged at molting. Together with old covers an insect is released from harmful products of vital activity which are accumulated in covers.

Each species undergoes certain number of molts, with this process being dependent upon length of larval development. Each growth stage of an insect between molts is referred to as an instar. If imago is formed after the last larval molt, it is called as development with semicomplete transformation. Under such development the larva is identical to imago and differs only by bodily forms and absence of wings, especially if it lives in the same medium as imago.

In highly developed representatives of insects class preimaginal development (before imago) takes place with complete transformation (metamorphosis). Under such development the last instar is changed into not imago formation but chrysalis (obtected pupa) during which insect organism is fully reorganized. Larval organs are transformed and formed again as imaginal organs. Such kind of development is typical for Coleopteran, Lepidoptera, Hymenoptera and Diptera.

Pupa can be open as is seen in beetles – some parts thereof are not covered by integument, they are close to each other. Such pupae coverings are chitinized, but they are usually thinner as compared with larvae and pupae are to be additionally protected. As a rule, this protection is in substrate: so, Colorado beetle pupa is developed inside the cavity in soil prepared by the larva. Butterfly pupa is protected by stiff integument which allows them to be placed openly, in Diptera pupa is hidden in additional cocoon i.e. puparium formed by sclerotized coverings of larva of the last instar. Insect pupae are very sensitive to any influences, despite they are protected. Under unfavorable conditions pupae may die, or they are wrongly developed. Due to this, adult insects appearing after imaginal molt experience different deviations in both body structure and in functionality of organs, ultimately resulting in their death.

In some cases prepupa is preceded by pupa formation: last instar larvae stop feed on and move actively up to pupation.

Imaginal development starts immediately after birth of an adult insect and is reduced principally to completion of its reproductive system and final development of other body systems. By its duration imago life in different species varies greatly: there are short-lived (ephemer) species which imago lives for 1-2 days even without feeding on (their mouthparts are not developed), but they manage to give birth to their generation (Ephemeroptera), and there are long-lived species (cockroaches, some beetles) which imago can live for several years. Periods of suspended development i.e. different forms of diapause significantly increase insect lifetime. As a rule, diapause is related to certain environmental changes, mostly to changing of daylight length and temperature. Different species have diapause in different instars. Along with it, there is both a mandatory diapause, which is to be passed through by the insect either to finish its development or to start reproduction actively, and non-mandatory (optional) diapause when conditions unfavorable for insect life appear. This diapause does not occur under optimum keeping of temperature, humidity and duration of daylight hours. Insect culture in both industrial and laboratory scale is based upon it.

Insect reproduction is a way for preservation of species compensating short lives and sensibility to influence of different unfavorable factors of some species, principally for the account of quite high fertility. Sexual reproduction occurs in most insects, however, in representatives of some orders (aphides) generations reproducing in sexual way are replaced by generations with asexual reproduction (parthenogenesis). Therewith even one female is able to be a stem mother for a new aphid colony.

There are organs allowing insects to be reproduced in sexual way, and to restore its population: e.g. Colorado beetle females have developed adnexa of generative chamber called spermatheca. After mating, sperm is received into the spermatheca and can be viable there for 8-9 months (winter diapause), and even for longer periods in some species.

Next summer such a female can lay up to 900 fertilized eggs giving birth to reproduction of new species.

Reproduction in social insects is regulated by complex system of hormonal and pheromone control. Female bee (fertilized female) is also able to keep sperm of males (male-bees) which it is copulated with for a long time, and under favorable conditions it can lay fertilized eggs (up to 2-3 thousand per day within the whole season of reproduction, i.e. 4-6 months).

Larvae which are further developed into either workers, or (under special conditions of feeding and development) queen bees i.e. large females forming new families around them, come from these eggs. Males (male bees) are developed from unfertilized eggs and therefore, if number of some species in one family sharply decreases, queen bee can easily restore number of male bees for further mating.

High fertility makes insect-pests more dangerous.

1.4. Mites - Biopolymers Pests

Mites belong to a separate arthropoda class. By their body composition they are closely related to the spiders – they have united body without separating into thorax and abdomen, and 8 legs. Mites are developed with partial transformation. One of their peculiarities is that small species as compared with other insects are not visible during the period of initial invasion.

Flour mite *Acarus ((Tyroglyphus) sirofaninae L.* can be found everywhere in the world, including Russia. It is one the most important and dangerous polyphagous pests living in different foodstuffs of both vegetable and animal origin. Basically, it is known as permanent pest of grain and products made of it.

Mites destroy grain corcule. Penetrating under the corcule they can develop there making latent form of infestation, so they are mostly dangerous for seed grain. They can be found also in cheese, vegetables, dried fruits, hay, straw and other plant residues. In the field flour mites are often nested with mouse-like rodents. It can also be seen in granaries of freshly harvested seeds.

The body length of mature adults of flour mites is up to 0.67 mm, it is oval and basically of albescent color and bright; head and legs are of henna-red or dark-violet color. Two pairs of bristles at the end of the body are longer than their length. Adult males of flour mite are smaller than females, their fore-legs are fusiformly thickened and possess aculeiform processes.

Optimum temperature for flour mite vital activity is 21-27°C, optimum relative air humidity is 80%, the minimum one is 65%. Mites cannot live in seeds with humidity of up to 13%. All life cycle at optimum temperature is 11-28 days. Flour mites are tolerable to low temperatures, however, they are sensitive to heating. At –5°C motile mites live within 18 days, and their eggs – 168 days. At 55-60°C mites die in 5-10 minutes.

In central parts of Russia flour mites can have 6-7 generations per year, in southern parts – over 10 in cold premises. Mites belong to a separate arthropoda class. Their bodies is similar with that of the spiders – they possess joint body without separation into thorax and abdomen, and 8 legs. Mites transformation is partial. One of their peculiarities is that small species are not always seen during primary invasion as compared with the insects.

Forage or food mites (Figure 13) (superfamily *Tyroglyphoidea*) are called so because there are many forms of them damaging both grain in granaries and other foodstuffs. By number of species Tyroglyphoidea are behind other species, but their abunance both in nature and in supplies of products is often extremely great. Tyroglyphoidea easily invade all possible organic matters and due to their polyphagy and high reproducibility within short period of time reach huge number. Unfavorable conditions result in death of active forms, but they do not affect hypopus stage which is non-feeding, can survive in dry environment and give rise to new mite colonies when get into suitable surrounding. It leads to their numerous and omnipresent dispersion in places with enough moisture and availability of organic matter. Structure of feeding phases and hypopuses is sharply different. Adult mite are usualy 0.3-0.5 mm in length, body is oval, covers are thin, lustrous, non-wetting due to adipose gland excretions opened along body sides. Coloring is whitish, fat-rich spare tissue is seen through covers. Legs and mouthparts are darker. The body is with permanent number of bristles which are of different size. Mites have no eyes. Main joints of short pedipulps are united into the blade, chelicerae with chewing claws. External genitals are modified due to special way of copulation. Under reproductive valves in females there is an egg-hatching pore, and male has a copultive organ. At mating the male is attaching behind the female at an angle, sperm comes to the copulative pore which is below female rectal orifice and connected with its reproductive tracts through tubules. Fertility varies greatly in different species, it depends on food taken by female and usually reaches 2-3 dozens of eggs, but sometimes exceeds 600. At higher humidity and excess of food eggs leaving out hypopus stage are immediately transformed into tritonymphs, and the latter – into adult mites.

Due to the fact that deutonymph stage (hypopus) in this case is left out, tritonymph in Tyroglyphoidea has been specially called as teleonymph. Tyroglyphoidea hypopi can be either resting or disseminating. The former are left in the matter waiting for favorable conditions. Such hypopi quite often have no appendages and look like egg or cyst in compact shell in their appearance. The disseminating hypopi possess clypeate body, strong shell, short legs, half-grown mouthparts and attaching claws.

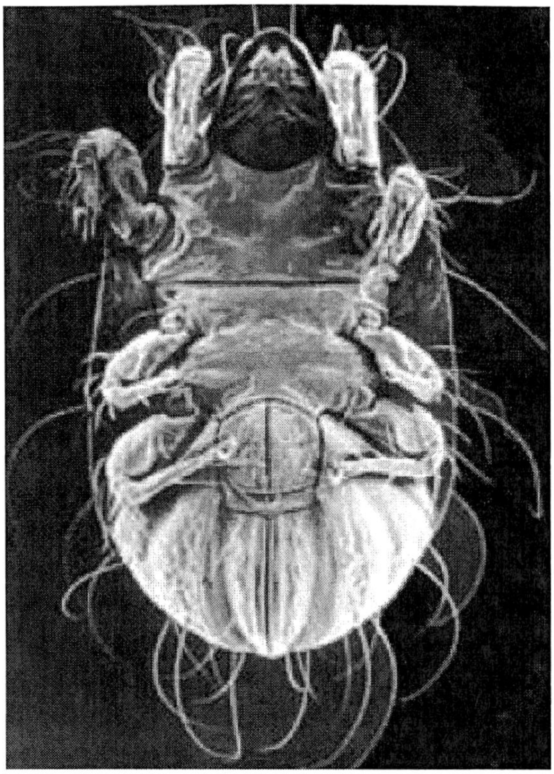

Figure 13. Food mite.

There are hypopi living in insects, and hypopi living in vertebrates (mostly rodents). The former possess attaching disks at the underside with several pairs of suckers – modified bristles. With their help hypopus is attaching to insect glabrate cuticle.

The latter ones instead of disk possess two pairs of chitinous claws clutching hair of an animal-carrier. Hypopi are dissiminating in those insects which are developed in mite-settled matters, and in accidental visitors. The are waiting for carriers on the matter surface and immediately attaching to moving object. Just bring a needle to them and these tiny creatures will stick to its tip. Species living in rodent nests and cache are dissiminating in them. Tyroglyphoidea hypopi living in ant colonies, termitaria, and bee-hives are dissiminating in appropriate insect hosts. Quite often over hundreds of hypopi can be found in insect body. Hypopi mechanism is not solved yet and has the most important general biological peculiarities. This phenomenon is principally similar to diapause in insects but it differs in that the insect diapause is due to usual cycle – egg, larva, pupa or imago, and in Tyroglyphoidea it is a special stage – hypopus. Hypopi are developed similar to diapause. In one case a number of generations takes no hypopus stage, then after environmental deterioration – aridity, lack of food, etc. – protonymphs are transformed not into teleonymphs but into hypopi, with this transformation being made very often later on, as soon as the environmental conditions are improving. In other species hypopi appear beforehand, thereby foreseeing unfavorable conditions for their transformation into teleonymphs. And there are species in which hypopus is developed in each generation. Hypopus endurance is striking and is due to hypometabolism in addition to shielded cover. Hypopi can endure dessication and freezeback, effect of poison and high dosage of ionizing radiation which is fatal for other organisms. Comparing life cycles of Tyroglyphoidea and Oribatidae it is easy to see that those functions of dissiminating and survival which are made by adult stage in Oribatidae, are performed by hypopus in Tyroglyphoidea. Therefore, Tyroglyphoidea adult stage has no imago, it is essentially propagative nymph in those cases when there is no hypopus and number of molts is reduced by one as compared with original cycle. Tyroglyphoidea are dispersed not in hypopus stage only. Some species with long bristles are gone with the wind to great heights. Tyroglyphoidea eggs ingested by rodents do not die but are dissiminating with excrements. The man facilitates these mites dissimination while carrying products. Tyroglyphoidea habitat is very different: soil, forest floor, clumps of plant residues, rotten wood, running sap of trees, fungi, lichens, mosses, roots, tubers and green parts of higher plants, nests of mammal, burds and insects, etc.

Characteristic species are found in grain-elevators and storage facilities where mites are carried from fields at harvesting. At grain humidity of 17% mites eat away corcule which is entered through damage of cover, and at higher humidity there is mite gradation in grain, they also feeding on endosperm. Apart from direct damage the mites affect grain germinating capacity, clog it with excrements and skins resulting in agglutination and spontaneous heating, infest it with putrescent microorganisms.

Mites are able to infest stored seeds and grain exhausting corcule: *Tyrophagus longior* mite feeds on seeds of wheat, oil-bearing crops and even stored yeast. It is also a serious glass-house culture pest. *Tyrophagus putrescentiae* mite affects sugar-beet seeds. *Suidasia nesbitti* causes damage to dried and smoked fish. Mites of *Tyrophagus*, *Carpoglyphus*, *Acarus*, *Glyciphagus*, *Cheyletus* genera are often found in cheese dairy storehouses. While examining meat and sausage plants, grain-elevators and large food warehouses 3 families of mites were discovered, with *Tyrophagus putrescentiae* being the most often in food stocks.

Flour is strongly affected by *Aleuroglyphus ovatus*. Cheese mite *Tyrolichus casei* is reproduced in cheese; *Carpoglyphus lactis* mite lives on substances containing lactic, acetic and succinic acids – dried fruit, sour milk, beer, rotten vegetables, etc. Mention should be made to *Histiogaster bacchus* mite which is sometimes mass-reproduced on the surface of wines in tanks of winery pouring shops and spoils wine. This genus species occur in nature in running yeastly sap of trees, especially of oaks. These mites appear to come to wineries with oak barrels used for wine transportation. Protection of grain against granary mites includes a number of measures. Due to dissiminating abilities of Tyroglyphoidea it is difficult to stop their getting to storehouses. Therefore, special attention is paid to storage conditions excluding mite reproduction: cooling of grain and maintaining its low humidity, elimination of self-humidifying and spontaneous heating factors, humid air access, etc. Chemical protection includes treatment with gaseous chloropicrin and powder-like poisons (acaricides), but application of these extraordinary measures is the result of bad storage conditions.

Besides, mites affect nonfood either, and apart from direct harm they induce allergy in people contacting them or affected products.

Listrophoridae (family *Myobiidae*) parasitize on small mammals. They stick to hair and suck out contents of hair follicle, lymph and blood plasma. Follicle mites (family *Demodicidae*) come from Myobiidae which live in sebaceous glands and bursa pili of mammals. They are small enough and can only be seen at high-magnification inspection.

They possess vermicular body, annulate cuticle, short legs as if chopped off, sucking mouthparts; they lay eggs; larvae, nymphs and adults are similar in appearance. Species of *Demodex* genus are parasites of different mammals, some of them are harmful for livestock farming causing follicular mange (demodicosis). Signs of infection become apparent in the cows three months later. In forebody skin there appear round bubbles of up to 1 cm in diameter. Sometimes these bubbles may have hair or be hairless. At breaks fluid of these bubbles stick hair together. After primary invasion into hair follicles of animals the mites insinuate into sebaceous glands cavity, encapsulate and later are found in capsules under the host skin as nodes (Kruglikovsky's spheres). Infestation takes place at direct contact. Reproductive life cycle of mite is as follows: egg, larva, 1st instar nymph (protonymph), 2nd instar nymph (deutonymph), imago (mature mite). The life cycle is 25 days, according to other data – 15 days. Many generations are over spring-summer period, which is due to higher insolation, changes of environmental temperature, immune and endocrine disorders.

Disease is hardly noticeable in sheep and goats, bubbles with mites can only be seen after shearing. Bubbles on the head – around eyes and on cheeks are found in short-wooled goats. Skin is flaking at strong affection.

After dehairing for offal demodicosis-affected parts of tanned unfinished products are seen as deep damages of glassy layer which significantly causes troubles in making high-quality leather.

Chapter 2

2. Characteristic of Main Insects-Pests

Insects injuring manufactured products and foodstuffs are not numerous but they can cause significant damage. They are representatives of five orders: cockroaches, hymenoptera (ants), termites, beetles (dermestids) and lepidopterans (moths).

Cockroaches can affect wool, leather, cotton-wool, paper, not mentioning foodstuffs, however, they can cause significant damage only at high abundance. Croton bug or German cockroach (*Blattella germanica)* is quite common in the territory of Russia. Its name is taken from point of view arising still in the 19th century that it was taken to Russia together with troops coming back from Germany after 7-year war. At the same time in Austria this cockroach was named as "Russian" considering it was brought there with Russian work-hands. German cockroach is often found in housing habitation, such species are called synanthropic.

For growth and reproduction it needs dark, well-heated (about 20°C) premises. Adult insect possesses well-developed wings of red-brown color, and body of 1-1.5 cm in length. Male differs from female by prominent appendages at the end of abdomen. Cockroach larvae are very small, with no wings and of darker color as compared with adults. German cockroach has 6 larval molts and 1 imaginal molt. Immediately after stripping off old skin (exuvium) larvae and young imago are white, covers are soft enough; they are becoming hard and and colored within several hours. Adult insects live up to 7 years. After mating female produce eggs in an ootheca (stiff chitinized cocoon), from which there appear larvae (nymphs) in several days. Cockroaches are developed without pupal stage, i.e. hemimetabolous development.

Cockroaches are very fast running along any (both horizontal and vertical) surface. They are reproduced the whole year round and hide in any chinks due to their flattened bodies. All this makes their control uneasy, besides, long-lasting attempts to affect them chemically make them chemoresistant. The most efficient way to control cockroaches are traps treated with new insecticides – pyrethroids and neonicotinoids, etc. To protect goods there are gels containing poisons which can be applied to any surfaces close to cockroach swarm. However, it should be remembered that regular aeration and reduction of temperature in premises can help to control these insects.

Termites injuring not only stored goods but wooden structures, and furniture are found principally in southern regions. In the Former Soviet Union they can be seen in Transcaucasia and Central Asia. These insects have features of social organization, making families of vast numbers (from several hundreds to millions), which depending on their functions are different in size, lifetime and behavior peculiarities, i.e. divided into castes. Their colonies include males, females, complemental kings, soldiers and workers. Termite control is the same as with ants – application of poisonous baits.

Ants are socially similar to termites but anyway differ strongly: they belong to Hymenoptera order which development is holometabolous. Caste division in numerous ant families is more complex. They are found actually on the whole territory of Russia, however, they are not always harmful to nonfood goods. Ants are more harmful to foodstuffs. The most dangerous species is Pharaoh's ant (*Monomorium faraonis).*

Minute size, capability to rapid reproduction, as well as ability to survive (up to 6-8 months) without food make ants one of the most difficult to destroy insect species (Figure 14). Settling everywhere in Russia this species has no natural enemies. It nests in heated premises in different places: in cavities of walls, bricks, under glazed tiles, in floor chinks, inside furniture, etc. Ant bites are quite painful. This species is suspected as carrier of bacteria and viruses, of poliomyelitis, in particular.

Population of this ant species is capable to significantly damage not only collections of insects by biting them and taking away their parts to nests, but insectories as well contaminating containers with cultures of other insects, living on their food, and killing their eggs, larvae, pupae and adult insects cultivated in cages. Principal control is application of poisonous baits.

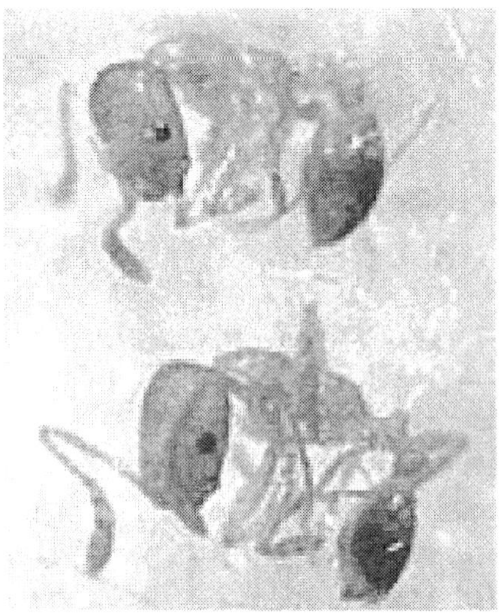

Figure 14. Pharaoh's Ant *Monomorium faraonis.*

Dermestidae belonging to beetle class (Coleoptera) are serious pests of storehouses and museum premises. These minute (up to 4-6 cm in length), dark (brown, brown-grey or black) beetles, with hard covers are often found near long-storing items of wool, fur and leather. Imago of the most common species – larder beetle *Dermestes lardarius*, fur beetles *Attagenus pellio*, museum beetle *Anthrenus museorum* – can feed on vegetable food, more often – pollen, however their larvae make more serious damage to stored fur, leather and wool.

Beetles (*Coleoptera*) from Dermestids (*Dermestidae*) class are the most typical pests of collections. Dust at the bottom of collection box indicates of pest presence (Figure 15). Dermestids larvae eat dry insects either fully or partially thereby making all life cycle at the feed source. Such population can exist for quite long time replacing several generations. Representatives of many genera of this family such as *Dermestes spp., Attagenus spp., Anthrenus spp* are also harmful for collection. Their larvae are well identified by tear-shaped form of a body and availability of long bristles as a "tail" at its end (Figure 16).

Some dermestids damage stored dried meat and smoked fish, are found in storehouses of cheese dairies. More often they are already mentioned larder beetles and related species (*Dermestes ater, Stegobium paniceum*).

Figure 15. Dust at the bottom of collection box indicates of pest presence.

Figure 16. Khapra beetle larva at the beginning and at the end of its development.

There are tobacco beetles (*Lasioderma serricorne*) amongst coleopterae. Molting skins found in goods and products more often indicate of dermestids presence. Larvae appear and develop not the whole year round but in warm summer months, therefore, storehouses, wardrobes and other storage places are to be examined more carefully in spring when imago appear in open places and windows. Heated premises can forward their all-year reproduction. Larvae of these beetles pass through several molts in their development (from 5 to 7) for 20-55 days and then make very mobile pupa covered by the last larvae skin. Larvae feed on fur, wool fibers, biting on leather items making them fall to pieces like sand. They do not only eat materials but bite through them long enough (5-10 cm) paths. Before pupation many larvae come out of the matter and damage therewith any materials and things found nearby.

Dermestids are basically controlled by premises fumigation, maintaining low temperature inside them and direct treatment of their clumps by appropriate insecticides. However, sometimes they can be of use, e.g. in some American museums skeletons are prepared with the help of dermestids larvae which are thoroughly gnawing round soft tissues and rigid tendons of articulations. Amongst representatives of Coleoptera order there are many pests of grain and other foodstuffs. The most typical ones are flour beetles, confused flour beetles and red flour beetles (*Tenebrio molitor, Tribolium castaneum, T. confusum*), grain weevils (*Sitophilus oryzae, S. granarius, S. zeamais*), cowpea weevils (*Callosobruchus maculatus, C. chinensis, Acanthoscelides obtectus*).

Saw-toothed grain beetle (*Oryzaephilus surinamensis*), khapra beetle (*Trogoderma granarium*) are often found in grain and flour reserves. Moth family (Lepidoptera order) is probably the most common typical species damaging clothes, fabrics, yarn and furs. The most common are webbing moth *Tineola furciferella*, naked clothes moth *Tineola biselliella*and case-bearing clothes moth *Tineola pellionella.*

Imago of these insects is small butterflies. The wings are covered with bright silver or yellowish scales. Pair of fore-wings is covering hindwings and the whole body of butterfly. Adult butterflies do not feed, their mouthparts (proboscis) are undeveloped.

Moth larvae are small caterpillars with a large number of pseudopods and well developed mouthparts (Figure 17). Caterpillar development in different moth species takes place in different periods, but damage they make is the same: they are able to digest basic protein occurring in fur and wool – keratin, they eat wool fibers, fur, floss and even synthetic fibers and fabrics. Caterpillars make thin cobwebs – indusia or tubular paths interweaving fiber residues.

Figure 17. Clothes moth larva.

These shelters are found more often in moth-spoiled things together with larvae excrements looking like fine sand. In their development caterpillars pass through 6-7 instars, then they pupate and after some time adult butterflies appear. They live 7-10 days, oviposting 60-100 eggs during their life time. Quite high temperature (above + 20°C) is necessary for the caterpillar development, therefore, aeration and cooling of premises where the goods are stored keep protect them from being spoilt by these insects.

Closewings found in food stores are similar to moths by their biological peculiarities. Their butterflies and caterpillars are larger than in moths and they are easy to see. European grain moth *Sitotroga cerealella* and rice moth *Corcyra cephalonica* are often found in grain and seed reserves, but most often dried and desiccated fruit stocks are to be protected against such closewings as almond and dried fruit moths. They are *Plodia interpunctella, Cadra cautella C. figuliella, Ephestia kuehniella, E. elutella.*

2.1. Larder Beetle *Dermestes lardarius l*

Occur throughout the world. In nature many larder beetles eat cadavers of dead animals. It is one of the most synanthropic species of dermestids (Figure 18). In living premises and storehouses it can breed on any products rich in animal proteins. In silk farming they destroy grain and strongly damage silkworm cocoons. Usually they eat dead or ill pupae *B. mori*, but they can also attack healthy ones. Larder beetle damage stores of dried and smoked fish. Adult beetles are about 7 mm in length. Optimum temperature for development is 18–20°C. Female oviposits 5–6 eggs within two months and their number can reach 175. Ovification lasts 7–8 days at 18–20°. At pupation larvae male tunnels (about 10 mm) in substrate, along with it they can damage wooden parts of furniture, premises, as well as items made of wool, fabric, books, etc.

Figure 18. Larder beetle.

Dermestids can switch from epizootic-killed dead animals to foodstuffs and carry agents of diseases. The researchers managed to find anthrax spores both on bodies and in excrements of three species of dermestids. Dermestids can be intermediate hosts of helminthes [4].

Attractant [5]

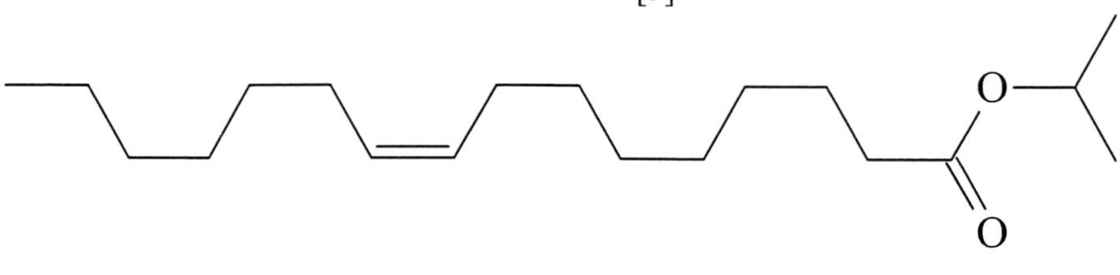

Isopropoxy-(9Z)-hexadecenoat.

2.2. Dermestid Beetle *Trogoderma glabrum herbst*

Occur throughout the world, in Russia found more frequently in southern regions. Dermestid beetle larvae damage grain and other phytogenous products. In nature they were found in nests of solitary bees. Beetles do not eat (Figure 19). Annotinous generation [6].

Sex pheromone components [7,8]:

(8*E*, 14*S*)-14-Methyl-8-hexadecenal

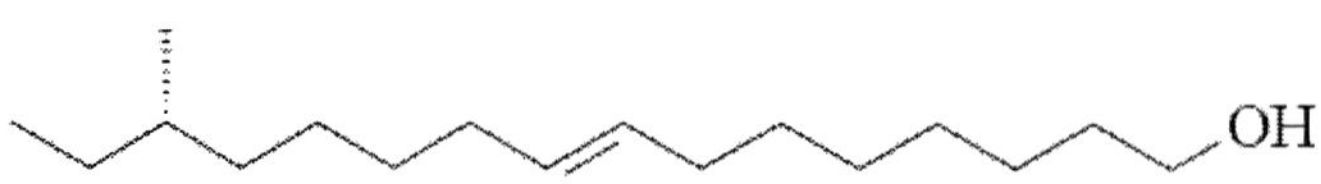

(8*E*, 14*S*)-14-Methyl-8-hexadecen-1-ale

Methyl-(Z)-7-hexadecenoat

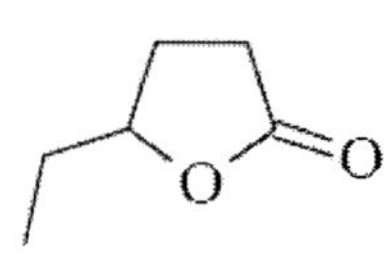

5-Ethyl-4,5-Dihydro-3*H*-furan-2-one

Attractants: octanoic acid, ethyl naphthalene, methylethyl naphthalene, tridecadiene, tetradecadiene-5-propyl-tridecan [9]

Figure 19. Dermestid beetle.

2.3. Khapra Beetle *Trogoderma granarium ev*

It belongs to the most dangerous pests of quarantine stores. The khapra beetle is acclimatized and found in storehouses of many world-wide countries, especially in subtropical belt and torrid zone, in Russia it is an object of external quarantine, there were cases of delivering with no stable foci.

It is larva which damages, being polyphage, and it feeds on dozens of kinds of phytogenous and animal origin products, preferring different grain-crops, readily eating seeds of leguminous plants, melons and gourds, flower crops, ornamental plants and forest plantations. This pest is more dangerous in dry and hot areas, as it feeds on grain with 6% humidity, while no other warehouse pests can feed on such grain.

The beetle is up to 3.2 mm in length; of tawny or hazel color; the top of the body is covered with thin slightly curved hairs of several colors (from Isabella color to black one), forming uncertain figure. Light hairs cover transparent ties on superior wings. Abdomen is covered with monochromatic, fawn hairs. Sexual dimorphism is clearly shown: males are less than females, their antennae with 5-joint club (females possess 4-joint one), the last sternum of abdomen is covered with back marginal, dark, half-adherent hairs going beyond sternum as fringe (Figure 20).

Eggs on one end are provided with several bisected spine-shaped outgrowths. Larvae are free-living of flexen coloring, legless, strongly wooly forming switches at the end of the body.The beetles neither fly, nor feed, they live from 5 to 30 days depending on temperature. Females lay by one egg, rarely by groups which catch rough surfaces (e.g. sacks) with their bisected spines. Low and high minimum effective temperature for khapra beetle are 15 and 46°C. Khapra beetle evolution does not actually depend on humidity of air and of food. At 32-36°C the development cycle lasts for 26 days, and at 27°C – 166 days. Overpopulation, lack of food and availability of large amount of excrements can initiate transition of larvae to diapause*, which can last over 4 years.

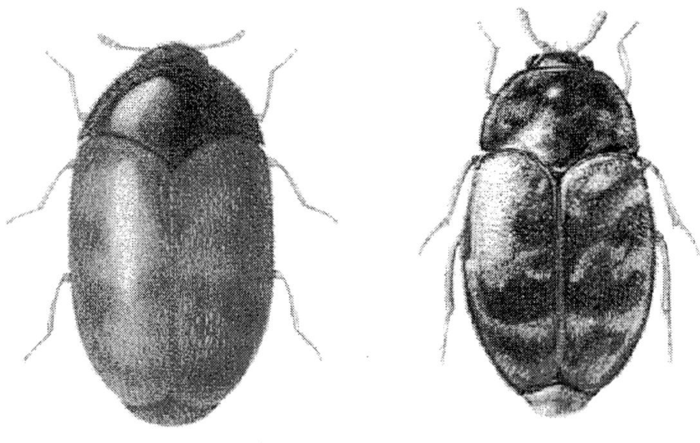

Figure 20. Khapra beetle.

Sex pheromones [10, 11]:

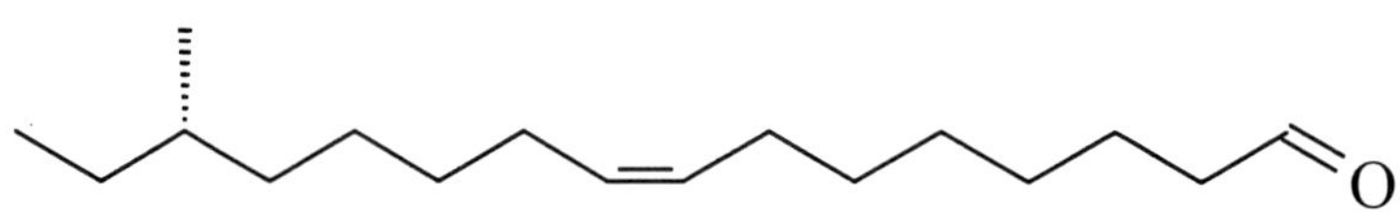

(8Z,14S)-14-Methyl-8-hexadecenal

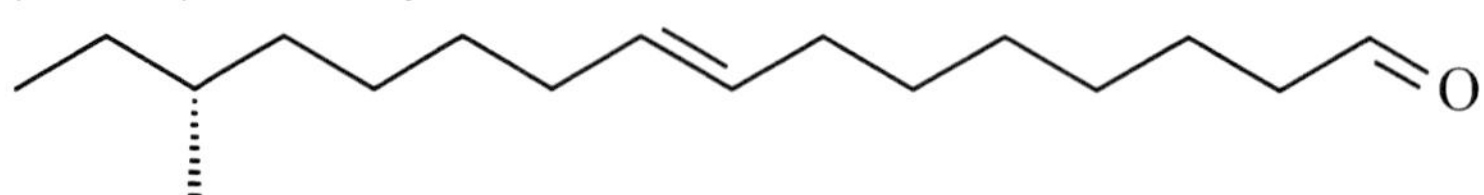

(8E,14S)-14- Methyl-8- hexadecenal.

2.4. Black Carpet Beetle *Attagenus unicolor unicolor brahm*

Cosmopolite. In nature it is usually found in bird nests [12]. Optimum temperature for development is 25°C. Black carpet beetle female lays up to 1656 eggs within 5–14 days. Ovification lasts for 5-24 days and depends on environment temperature. Larval stage development lasts for 1–3 years. It is one of the most wide spread and usual pests of keratin-containing biopolymers (wool, fur, etc.).

* Diapause – a period of suspended development and growth in living organisms. It usually takes place in winter

Sexual attractant [12]:

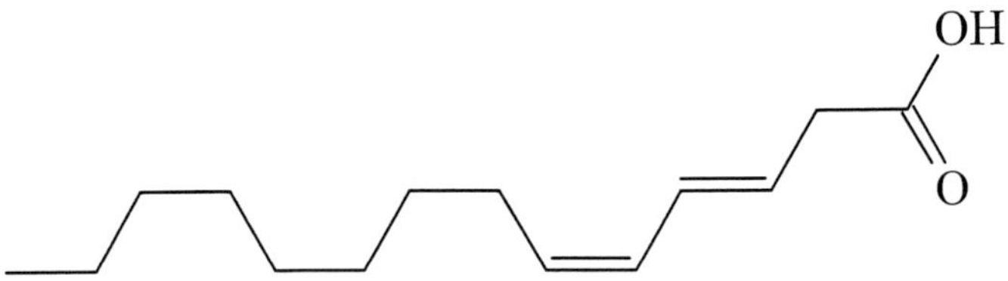

(3E, 5Z)- tetradecadienoic acid

2.5. Drugstore Beetle *Stegobium paniceum l*

Drugstore beetle occur world wide (including territories of Russia and the CIS). It is found in living premises, libraries, archival depositories, stores, bakeries, cereals factories and warehouses, rarely in granaries. The beetles do not eat, but they can make holes in dry food. Larvae damage different bakery foods, cereals and other products from processing of grain, they were found in the upper layer of long-stored grain.

The beetle is 1.5-3.5 mm in length, of foxy or oxide-yellow color, dull, covered with hairs. The body is short and of cylinder form, pronotum covers head like hood, it possesses sharp sides, its superior wings are with clearly seen dotty grooves (Figure 21).

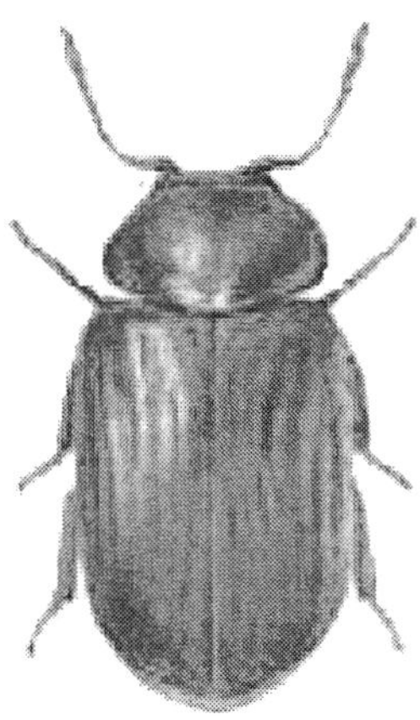

Figure 21. Drugstore beetle.

Drugstore beetles are active in the evening or at night. Light-oriented. They do not need additional food for reproduction. Males soon die after mating, and females oviposit and live from 2-3 weeks to one month. Females lay their eggs either on the surface of foodstuffs or on internal holes made by beetles by one or by groups. At 26-27°C one generation is developed within 66 days on the average, at 20-22°C – 219 days. In heated premises there can be found up to four generations of beetles and mass appearance is seen during the whole year. In unheated premises there are up to two generations, and mass flight of the beetles takes place in June only. After appearance of the beetles the foodstuffs are marked with numerous small

periods for survival under unfavorable conditions.

holes of up to 2 mm in diameter. At low moisture of foodstuffs (6% and lower) larvae intrude into substratum and make holes inside. In this case if food is infected development of drugstore beetle generation takes place inside it, and the beetles do not appear outside. Diagnostics of food infestation by drugstore beetles is difficult to be made in such cases. At high level of infestation the beetles go out of substratum searching for new habitats, and then they can be found on the surface of food.

Drugstore beetle Sex pheromones [13, 14, 15]"

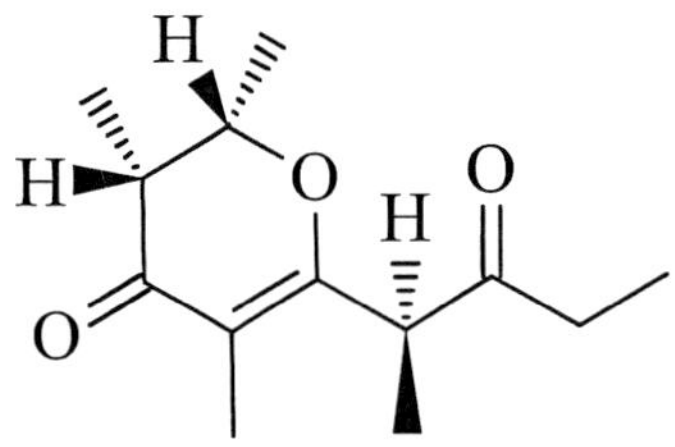

(2S,3R,1'R)-2,3-Dihydro-2,3,5-trimethyl-6-(1'-methyl-2'-oxobutyl)-4H-pyran-4-one

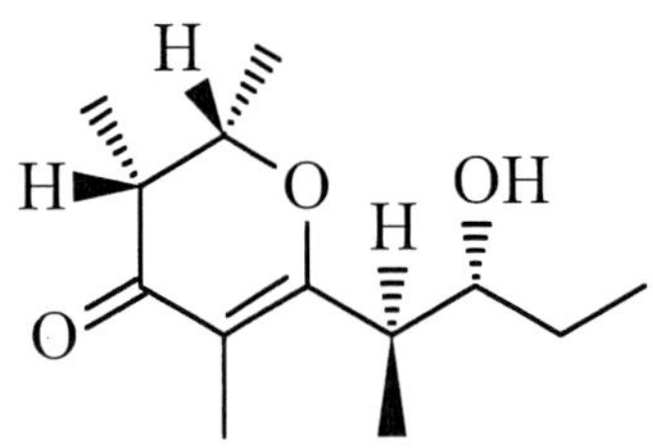

(2S,3R,1'S,2'S)-2,3-Dihydro-2,3,5-trimethyl-6-(1'-methyl-2'-hydrooxobutyl)-4H-pyran-4-one

2.6. Larger Grain *Borer Prostephanus truncates horn*

Larger grain borer was initially spread in Central America and Mexico only, where it was found in forests (it can develop on many kinds of wood), it was also marked long ago as local pest of stored corn. At the end of 70s of the last century it was occasionally taken to eastern and later to western Africa. Its spreading was quick and nowadays it is registered in all 13 African countries. In majority of them it became one of the most serious pests of stored corn and dry Brazilian arrowroot. In 3-6 months after storage it can destroy up to 34% of gathered crop. Larger grain borer is also found in some southern states of the USA, but no traces are found in other countries (Figure 22). It is beetles and larvae which cause damage. The beetles fly well. Female lays its eggs to corn ears and grain. All life cycle lasts for 25 days in central Africa.

Figure 22. Larger grain borer.

Aggregation pheromones [16, 17]:

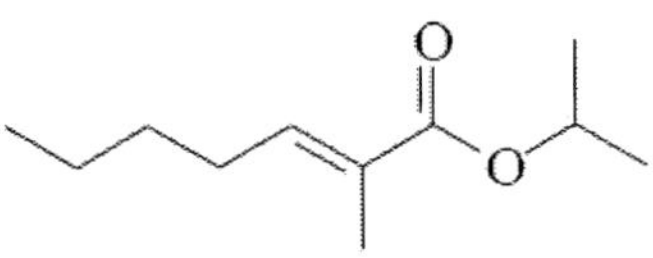

Isopopyl-(*E*)-2-methyl-2-heptenoat

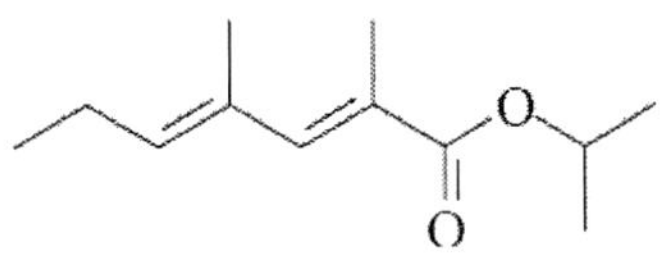

Isopopyl-(*E,E*)-2,4-dimethyl-2,4-heptadienoat

2.7. Lesser Grain Borer *Rhyzopertha dominica f*

Cosmopolite (mainly in tropical zones). In Russia it is found in southern regions. It inhabits warm premises of granaries, although they can be seen in nature. It belongs to the most dangerous grain pests damaging grains of corn, rice, barley, wheat, rye, oats, sorghum, buckwheat, as well as combined feed, cereals and seeds of green crop. Usually, the borer does not damage sunflower seeds and been cultures. It is the beetle and its larvae of the lesser grain borer that feed on strongly, thus resulting in thin gnawed case left from grain. Larvae and beetles excrete much. At strong infestation grain mass has honey-musty flavor.

The beetle is up to 3 mm in length. The body is elongated of cylinder form, bright, brown with umber or dark-cherry colorings. Pronotum covers head like hood, in the front it is covered with concentric row of denticles; elytrons have lengthway rows of dotty fossae; the last (5th) abdomen segment is yellow from below in females and fully brown in males.

Imagines* are active at any time of the day and night, but they fly in twilight. For reproduction they need additional food. Beetles of the lesser grain borer are able to survive at

* Imago (from Latin *imitar* – image (likeness) – the final adult sexually mature form of anthropoda with complex life cycle.

favorable conditions within several months and even up to one year. Female of the lesser grain borer lays its eggs onto grain either by one or by groups. Larva is developed within the grain. Optimum temperature for the lesser grain borer activity is within 32-35°C, with optimum relative humidity being 50%. Larvae can develop in grain with 8% humidity and at 37°C. Lower temperature threshold for the lesser grain borer development is 16.4°C. Temperature changing from 20 to 30°C, the life cycle of the borer varies from 11 to 29 days. Up to 8 generations of the pest can develop in warehouses.

The lesser grain borer is very sensitive to low temperatures and dies soon after the grain cooling, but it is very resistant to high temperatures, so it is difficult to destroy by simple grain heating, e.g. grain drying.

As a result of much feeding the beetles of the lesser grain borer often gnaw into corn seed and they are sometimes difficult to find in grain probes at seeding.

Aggregation pheromones: dominicalures 1 and 2 [18, 19, 20]:

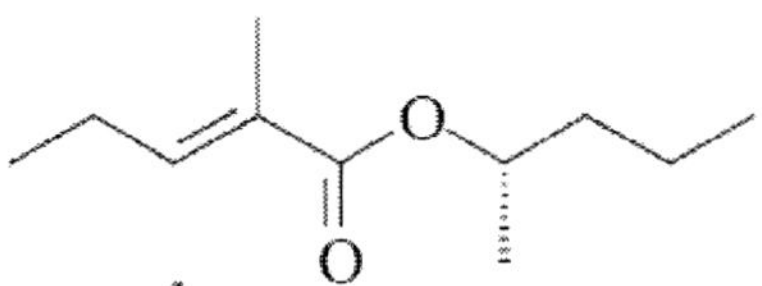

1

(*S*)-1-Methylbutyl-(*E*)-2-methyl-2-pentenoat

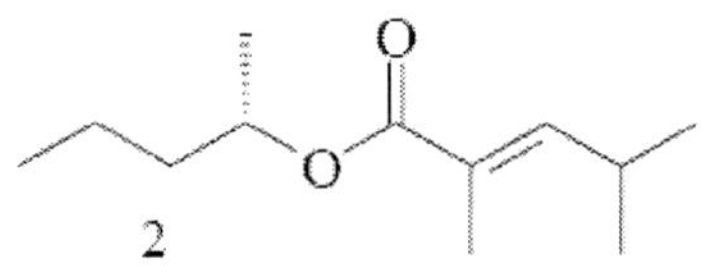

2

(*S*)-1-Methylbutyl-(*E*)-2,4-dimethyl-2-pentenoat

2.8. Red Grain Beetle *Cryptolestes ferrugineus steph*

Cosmopolite, in Russia it is mainly distributed in southern parts. Red grain beetle is found in warehouses, mills, cereal factories and bakeries, confectionaries and in storehouses with dried vegetables. In nature it can be seen under the bark of dry trees and in bark beetle holes.

It feeds on damaged grain and products of its processing, cotton seeds, cotton cake, cocoa beans, dried fruit, etc. Among grain crops it prefers wheat and rye, but usually it develops on grain already damaged by other pests. In many cases red grain beetle is found in cereals together with maize and granary weevils, beetles and some species of cucujid beetles. Gathering in warm parts of grain mass it forms colonies which result in spontaneous heating.

The beetle is 1.5-2.5 mm in length. Its body is flattened and of hazel color; antenna length in females is less than half of its body, and that of males is a little longer than half of its body; when turning back antennae do not reach abdomen end. Apical antenna joint is two times longer than its width. Length of pronotum equals to its width (Figure 23). The beetles are able to fly, especially in warm summer days. Imago of red grain beetle at 28°C and relative humidity of 75% can live up to 220 days. For reproduction they need additional food. Females lay their eggs either by one or by small groups. Among cucujid beetles found in storehouses red grain beetle is the most thermophile species. Lower threshold of its development is 18.5°C. At 28°C and relative humidity of 75% full life cycle of the beetle lasts 24 days on the average, if the temperature falls to 20° and below one generation develops over 200 days. It is very sustainable to lower temperatures and can give up to 4-6 generations in the south. It is difficult to find in grain probes at seeding.

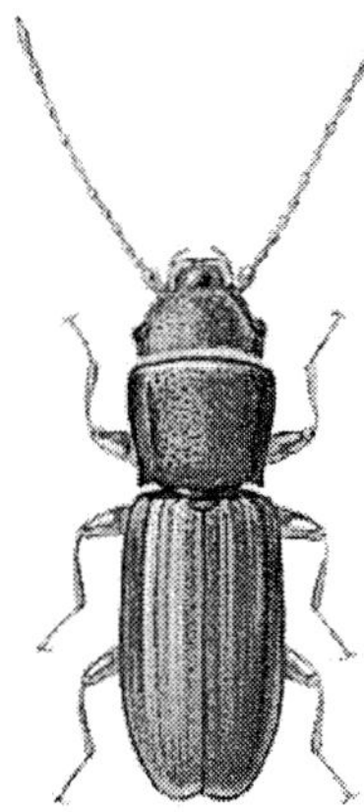

Figure 23. Red grain beetle.

Kairomones [21]

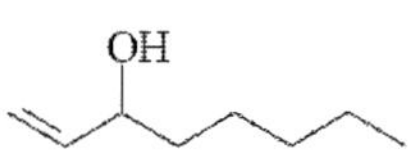

1-Octen-3-ole

3-Methylbutan-1-ole

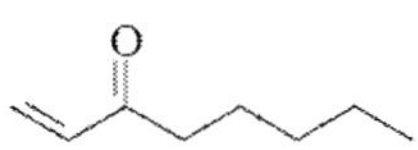

1-Octen-3-one

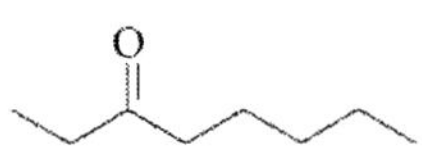

Octan-3-one

Aggregation pheromones [22]:

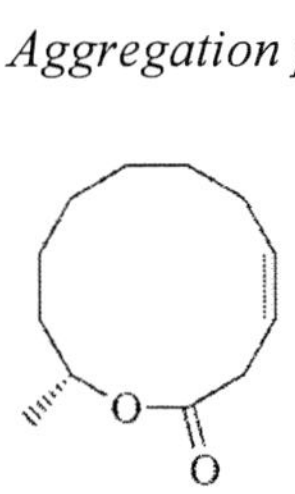

(3*Z*,11*S*)-Dodecen-11-olid

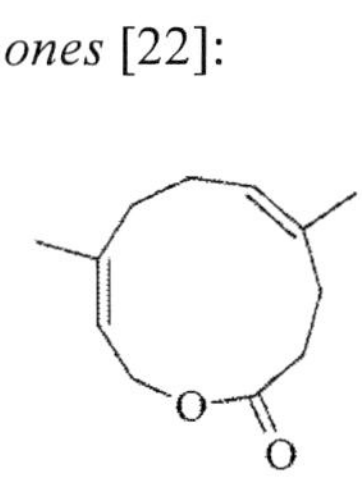

(*E*,*E*)-4,8-Dimethyl-4,8-decadien-10-olid

2.9. Merchant Grain Beetle *Oryzaephilus mercator fauv*

Cosmopolite. They are found in mills, confectionaries and macaroni factories, it prefers seeds of oil-bearing crops and nuts especially damaged ones. The beetle is 2.0-3.5 mm in length. By its external appearance it looks like saw-toothed grain beetle, however the latter one possesses larger eyes, the eye diameter is larger than temple area behind the eye, head is rectangular, protonum outgrowths are narrow and straight, and coloring of the body is usually darker (Figure 24).

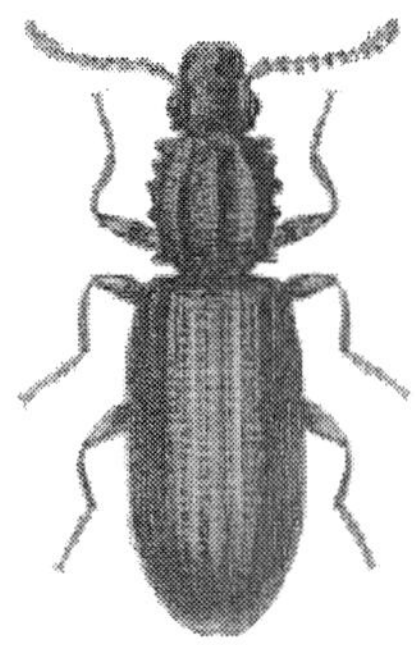

Figure 24. Merchant grain beetle.

Imagoes flight well, and live up to 6-10 months, and sometimes much longer. At 32-35°C development of life cycle lasts for 20 days. Main biological peculiarities are like in saw-toothed grain beetle.

Kairomones [23, 24, 25]:

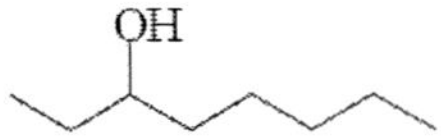

Octan-3-ole

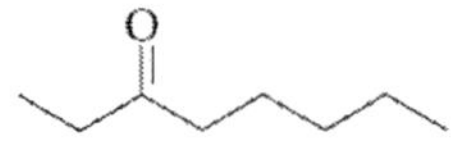

Octan-3-one

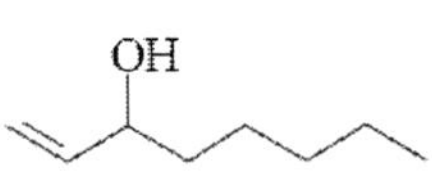

1-Octen-3-ole

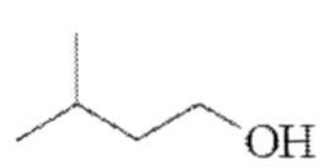

3-Methylbutan-1-ole

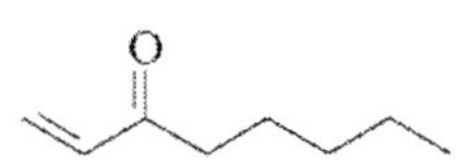

1-Octen-3-one

Aggregation pheromones [26]:

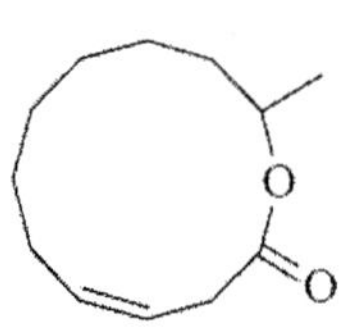

(*Z*)-3-Dodecen-11-olid

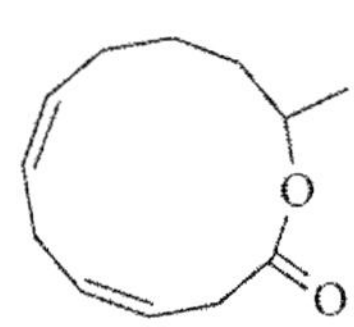

(*Z*,*Z*)-3,6-Dodecadien-11-olid

2.10. Saw-Toothed Grain Beetle *Oryzaephilus surinemensis*

Cosmopolite, it is widely distributed in Russia. It is found in warehouses, mills, confectionaries and macaroni factories, stores and living premises; in nature it can be seen under old trees bark. Grain products of all kinds, such as flour, groats, cereals, flakes, dough, bakery foods, etc. can be used as food for adult insects and larvae. When attacking wheat, rye and corn grains saw-toothed grain beetle first eats away seed-buds and then its endosperm. However, most of all it develops in grain already damaged by other pests. In nature they are predatory killing small insects (bark beetles, etc.) living under the bark of old, withered and defective trees. The beetle is 2.2-3.5 mm in length. Its body is narrow, dark-brown with dull luster. Temples behind the eyes are rounded, eyes are small. Prothorax is separated from abdomen by sharp constriction, pronotum has six large beaks on the sides and two lengthway grooves from above.

The beetles cannot fly, and live usually 6-10 months, but sometimes can survive within 2-3 years. The most optimum temperature for the saw-toothed grain beetle is 31-34°C, and relative humidity is about 65%. If air is dry (5-30%) the development period is increasing. Lower temperature threshold for development is 15.6°C. Average life cycle for the development at 20, 25 and 30°C is 78, 36 and 24 days accordingly. The beetles winter in unheated premises. As compared with other species of cucujid beetles the saw-toothed grain beetle is less cold-resistant. At 0°C it dies out in 22 days, at –5°C in 13 days. The pest can give 2-3 generations in the central part of Russia, and up to 4-5 generations and more in the south.

Kairomones [27, 28]:

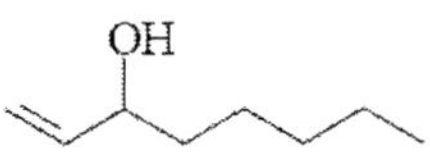

Oct-1-en-3-ole

3-Methylbutan-1-ole

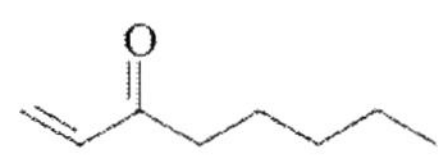

Oct-1-en-3-one

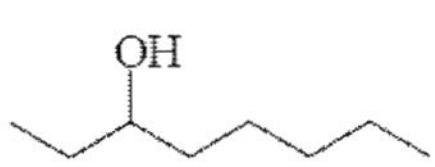

Octan-3-ole

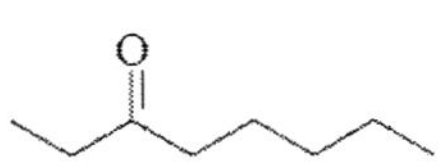

Octan-3-one

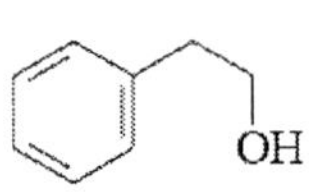

2-Phenylethanol

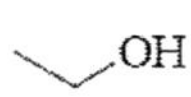

Ethanol

Sex pheromones [29]:

(Z,Z)-3,6-Dodecadien-11-olid

(Z,Z)-3,6- Dodecadienolid

(Z,Z)-5,8-Tetradecadien-13-olid

2.11. Yellow Mealworm Beetle *Tenebrio Molitor l.*

Occurs world wide, Russia including. It is found in mills, granaries, confectionaries and macaroni factories, bakeries, rural living premises. Beetles and larvae damage grain, flour, cereals, bran, mixed fodders, different flour products, cotton-plant seeds and cotton cake, tobacco, insect collections. Mostly they prefer grain milling products, e.g. flour, especially raw and musty. In normal grains the larvae first eat away seed-buds and then floury portions with glumes. Yellow mealworm beetle larvae are often called "flour weevils".

The beetles are up to 16 mm in length. The body is elongated and flattened, mostly of dark-brown coloring, dull glistening, the abdomen is dark-brown from below Sheaths are with parallel edges, their length is two times more than its width (Figure 25).

Figure 25. Yellow mealworm beetle.

The beetles fly well especially in twilight and at night, if touched they stink. Females lays their eggs either by one or by small groups on flour, bran and other products, and if they are not available – into the openings in walls, hutches, onto the sack surface. At 25°C the beetles live about 100 days, females live longer than males. Larvae can survive without food for a long time (up to eight months), they can resist lower temperatures. Development of one generation at 25-27°C lasts up to 225 days. In unheated premises the pest has usually one generation, in heated ones – up to two generations.

Sex pheromone:

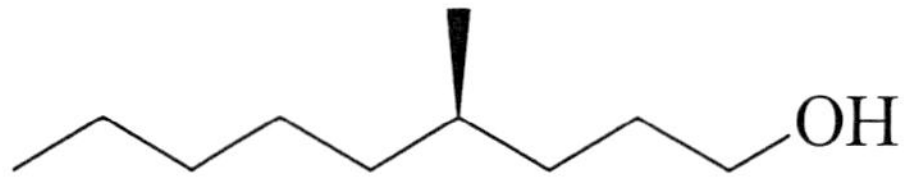

4R-Methylnonane-1-ole

2.12. Red Flour Beetle *Tribolium castanuem herbst*

Cosmopolite, occurs in all parts of Russia. It is one of the most dangerous pests of stored products. It is found in mills, flour storehouses, cereal factories and bakeries. It attacks flour, cereals, bran, mixed fodder, flour products, dried fruits, broken oil seeds and seeds of flax, peanuts, kidney beans, peas, and cocoa. It damages more broken grains that the whole ones. By damage it is far behind the confused flour beetle.

The beetle is 2.3-4.4 mm in length, its body is from red-brown to dark-brown color; the distance between the eyes from behind the head is no more than eye diameter; the three last antenna joints are much wider than the previous ones, forming the club.

Red flour beetle flies well, especially at night. The beetles have stink glands on abdomen and thorax segments releasing liquid with pungent acrid odour. Imagoes live up to 2 years and more, 220-540 days on the average.

At optimum temperature (30°C) full life cycle from an egg to adult insect lasts for about a month. Lower temperature threshold for the red flour beetle is 15.2°C. It is quite tolerant to lower moisture content, it can develop in в ground products at 1% of moisture content. Cannibalism is typical in the population. In heated premises they develop during the whole year round, in unheated premises it gives 2-4 generations.

Aggregation pheromone [30,31]:

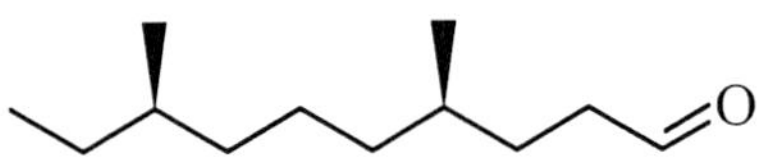

4R,8R-Dimethyldecanal

2.13. Confused Flour Beetle *Tribolium confusum duv*

Cosmopolite. It is one of the most dangerous pests of stored products. Being thermophile species, the confused flour beetle occurs more often in heated premises and grain processing enterprises, they make much harm in flour mills.

Beetles and larvae damage more often grain processing products, as well as peanuts, cocoa beans, tobacco, rarely – dried vegetables and fruits, they are able to attack whole wheat grains with humidity of 12% and higher.

In mills and cereal plants the confused flour beetle prefers to hide inside the equipment, behind the skirting boards and in other places where there are flour spillage.

The confused flour beetle is often confused with the red flour beetle. The confused flour beetle antennae are thickened gradually to the top, the eyes are less, in the lower part the distance between the eyes is three times more than the eyes width, the wings do not function, the body is 3.5 mm at length (Figure 26) [32]. The beetle needs additional food for mating and oviposition. At 20-25°C and normal feeding adult beetle life cycle lasts from 14 months to 3 years.

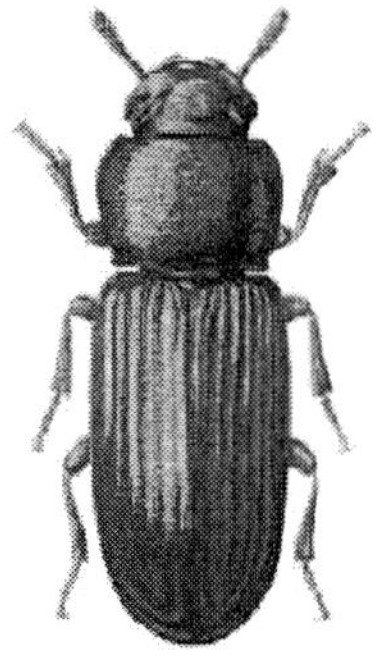

Figure 26. Confused flour beetle.

Lower temperature threshold for the development of the confused flour beetle is 14.8°C. With increasing the temperature from 20 to 30°C, the development life cycle decreases from 109 to 37 days accordingly, on the average. Optimum temperature for the development is 23-25°C, in this case the full development cycle lasts for 1.5 months. In unheated premises the beetle gives 2-3 generation in summer periods, and in heated premises there can be up to 10 generations.

Pheromones (allomones) [33, 34]:

Pentadec-1-ene

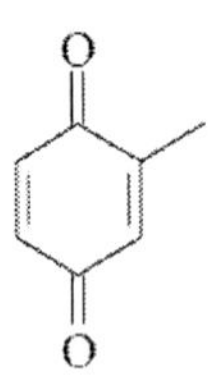

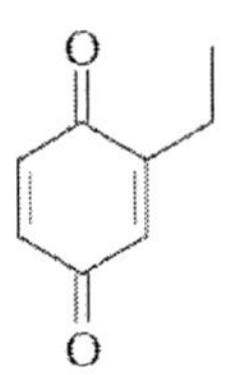

Aggregation pheromones [35]:

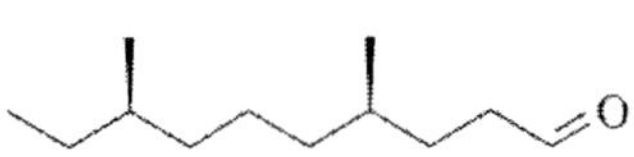

(4*R*,8*R*)-4,8-Dimethyldecanal

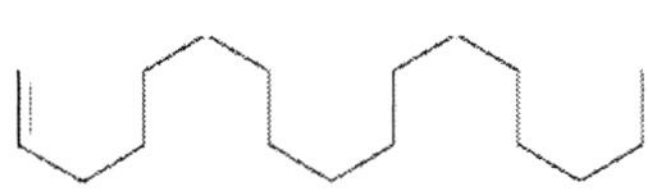

Heptadec-1-ene

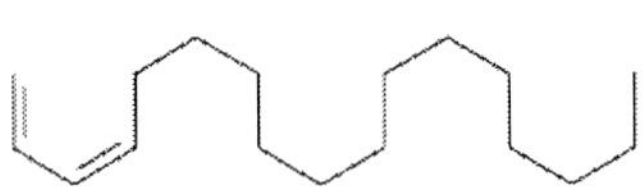

Heptadeca-1,3-diene

2.14. Dark Flour Beetle *Tribolium destructor vyttenb*

Occurs world wide, Russia including, especially in southern regions. Larvae and beetles are polyphagous. They damage grain, flour, bran, cereals, bakery products. In seeds of wheat and rye they eat away seed-buds. The beetles make the products malodorous reek of cresol.

The beetle is 5.0-5.5 mm in length. Elytrons are of dark-brown to black color, with abdomen and legs being brighter. The 4th and 5th antenna joints are gradually thickening to the top. Prothorax has strongly protrudent rounded edges (Figure 27).

Imagoes live up to three years. Favorable temperature for the development is 25–28°C, with the product moisture being about 15%. Under such conditions one generation develops for 45-60 days. Lower temperature threshold for the development is 13-14°C, with the upper one being a bit higher than 30°C. They are sensitive to cold, especially to sharp temperature drops, and at 0°C they die out in 19 days, and at –5°C – in 5 days. In heated premises they give 3-4 generations.

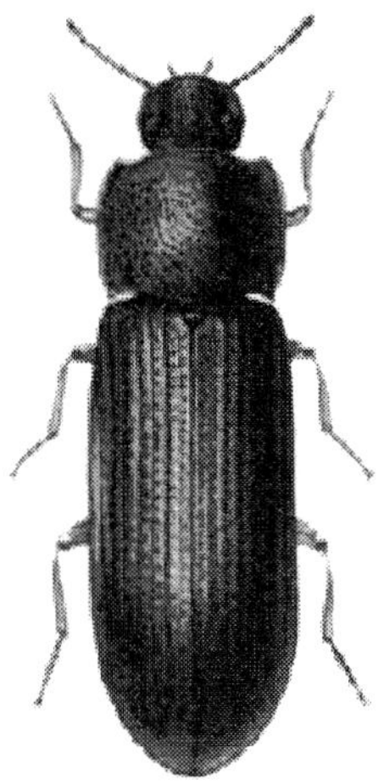

Figure 27. Dark flour beetle.

Aggregation pheromones [36]:

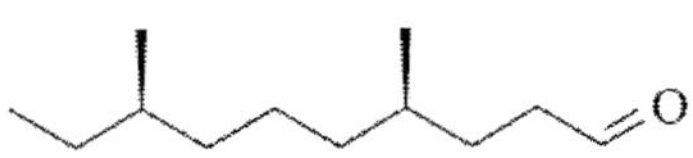

(4R,8R)-4,8-Dimethyldecanal

Pentadec-1-ene

Pentadeca-1,6-diene

2.15. Bean Weevil *Acanthoscelides obtectus say*

Occurs world wide in warm areas where the winter temperature rarely falls below 2°C, it was brought with kidney beans to many countries, Russian including (Northern Caucasia, Krasnodar territory). The weevils can be reproduced both in fields and in storehouses. Bean weevil larvae attack first of all different varieties of kidney beans, as well as seeds of soya, bean, pea, chick-pea, peavine, and lentils. Edibility characteristics and sowing qualities of seeds are reduced in damaged seeds.

Bean weevils are 2-5 mm in length; females are twice as much than males. The body is short and oval, densely covered with grey or drab hairs. Antennae at the bottom and on the top, as well as upper edge of elytrons are of reddish color, abdomen is of copper color on the top. It possesses 11-joint shortened antennae, when turned back, they hardly reach elytron base center. There are up to 10 dotted lengthway grooves and by 5 smudgy black spots. Hind femurs on internal side before the top are with large sharp dens and two additional denticles. 1st instar larvae possess well developed legs, larvae of next instars are legless, the whole development takes place inside the grain (Figure 28).

Wintered weevils are awakened after chill coma at 15-16°C, mating and oviposition starts in warm conditions, when ambient temperature exceeds 18°C. In cloudy days they are indifferent to light, and on sunny days they fly in its direction. On warm sunny days they are able to quickly run and fly well for long distances. Additional food does not seem to be required for imago. Females start oviposition in 1-2 days after mating. Bean weevil chooses beans with dry shells only. The female lays by 20-40 eggs to one place (between two seeds inside the shell, or to the surface of the bean seed). Hatched larvae bore inside the grain and develop there. Over 50 larvae can be found in one grain. In storehouses the female lays its eggs to bean seeds or sacks by dozens.

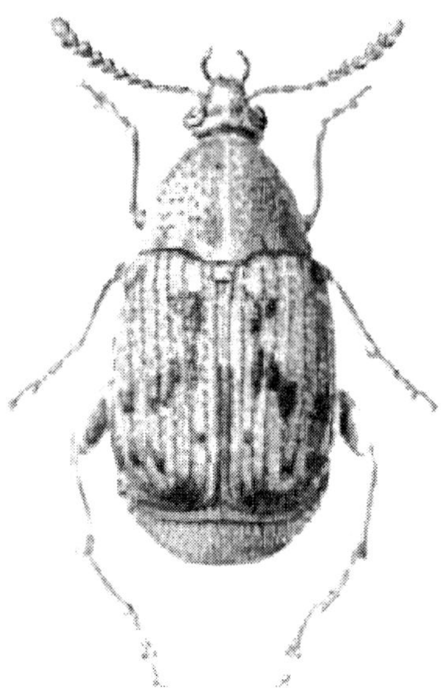

Figure 28. Bean weevil.

Optimum conditions for bean weevil development are: 27-31°C and relative humidity of 65%. In this case the full life cycle is 1.5-2 months. In the South it gives 3–4 generations.

Sex pheromone [37]:

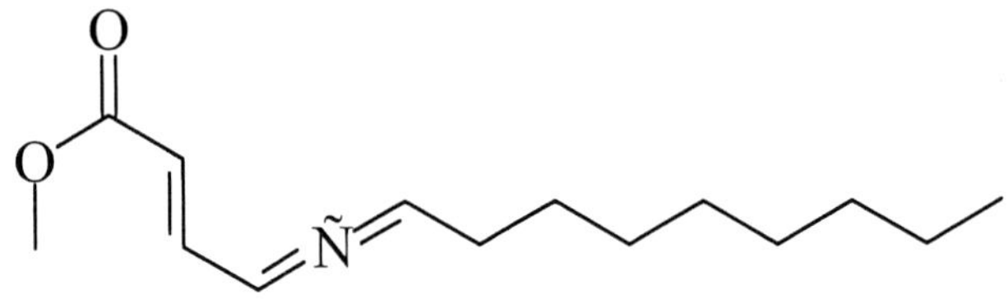

Methyl ether (*E*)-2,4,5-tetradecatrienoic acid

2.16. Chinese Weevil *Callosobruchus chinensis l.*

Occurs world wide in all continents, especially in southern regions. In Russia it is an object of external quarantine. It damages wide spectrum of leguminous plants, including pegion pea, field pea, field bean, chick-pea, soya-bean, lentil, and peavine in storehouses.

They are 3.5 mm in length, the body is short and of oval, almost rectangular form. Male antennae are long and comb-like, female has saw-like antennae. Elytons are of hazel color, however, coloring power varies greatly. Hind-leg femurs are thickened with long groove for placing shanks. Pygidium* is steep, densely wooly, light with attaching hairs. There are spots of light attaching hairs on the sides on 2-4 sternites (Figure 29).

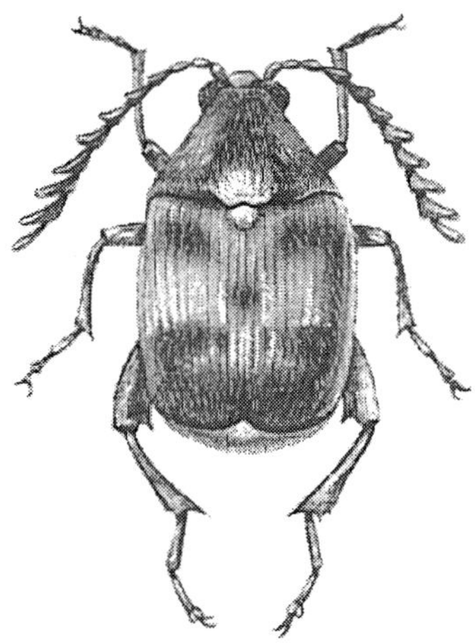

Figure 29. Chinese weevil.

Weevils leaving the seeds do not need additional food. They usually mate at the same day and the female starts laying eggs. The eggs are fixed to grain surface by special secretion. 1st instar larvae tunnel inside the seed where the whole development of the pest takes place. Each grain contains up to 10 larvae.

* Pygidium – terminal segment of the insect abdomen.

The adult weevils live about 12 days, rarely up to 36 days. Development from the egg to the weevil at 30°C lasts about 18 days, at 18°C for 40-48 days. The life cycle prolongs for 3-4 months in winter in storehouses. The activity stops at temperature below 10°C. Depending on conditions there can appear up to 6-8 generation within a year.

Sex pheromone:

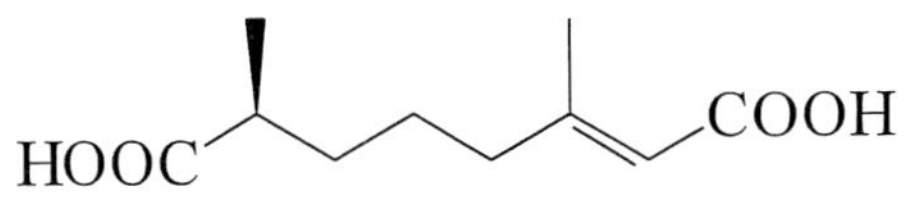

(2*E*,7*S*)-3,7-Dimethyl-2-octen-1,8-dioic acid

2.17. Cowpea Weevil *Callosodobruchus Maculates l.*

Actually it occurs world wide, preferring southern regions. In Russia it is an object of external quarantine. Cowpea weevil mainly damages in storehouses, but it can develop in the field either. Damaged cultures include vetch, pea, horse bean, mung bean, soya-bean, kidney bean, lentil, peavine, pigeon pea, dolichos and wisteria.

Cowpea weevils are 2.4-3.8 mm in length; the body is of puce color, elytrons are with black spots forming cross-shaped figure (Figure 30).

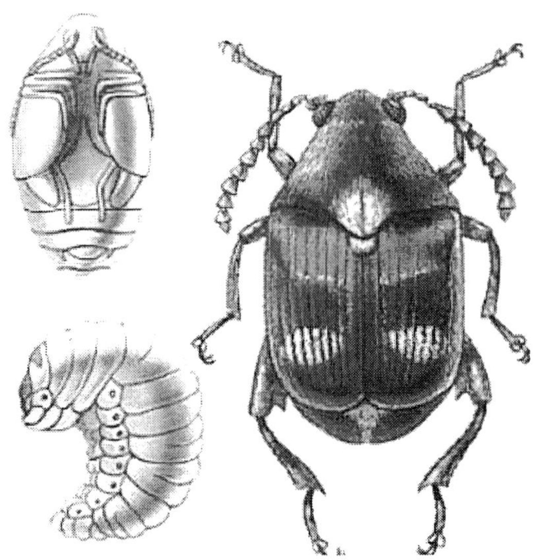

Figure 30. Cowpea weevil.

In one day after appearance the weevils start mating and oviposition. They are able to flit, especially in hot season, however there are species which cannot fly. Cowpea weevil females lay their eggs (from 2 to 15) on grains of growing beans or on seeds, as well as on sacks and sides of different containers. The eggs are strongly sticked by rapid-hardening secretion. Like other weevil species 1st instar larva tunnels inside the grain and develops there.

The whole life cycle varies from 40 to 180 days. At higher temperature (27-30°C) up to nine generations can develop.

Sex pheromones:

(maximum attractivity for compound E- and Z-isomers in ratio 50:50 or 25:75) [38]:

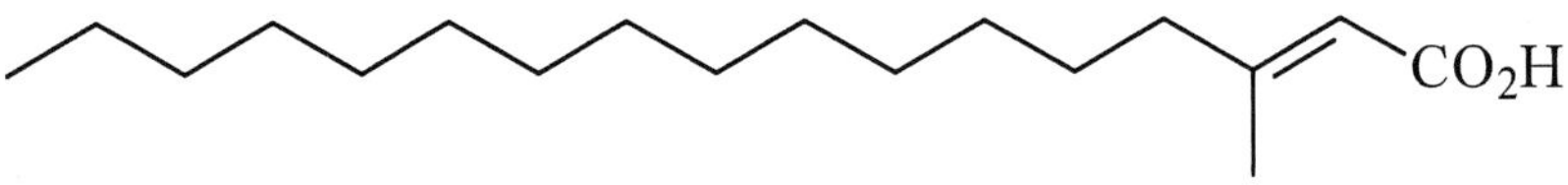

(E)-3-Methyl-2- heptadecenoic acid

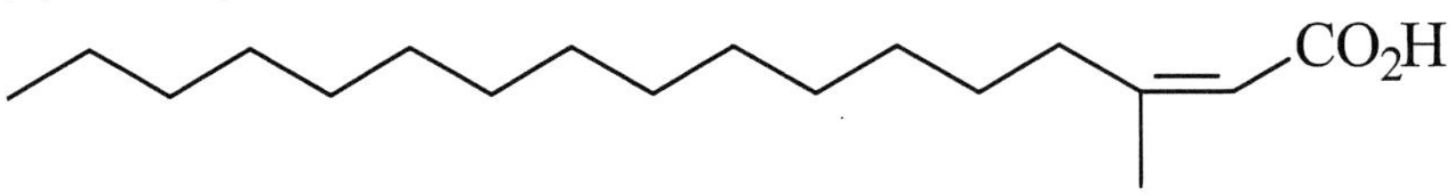

(Z)-3-Methyl-2- heptadecenoic acid

2.18. Granary Weevil *Sitophilus granaries l.*

Granary weevil is one of widely distributed and dangerous pests for stored grain. At present it is found world wide in moderate countries with mild winters; it is quite rare in Russia than the maize weevil. It is more often found in granaries and storehouses with bad ventilation.

It is larvae and weevils which damage. The most suitable food for the granary weevil larva is the floury tissue within the seed (endosperm), which is damaged by the weevil leaving embryo untouched. Among different stored grain the granary weevil more severely damages compact flour, pearl barley, rice grits, buckwheat, grains of wheat, rye and barley; less affecting grains of chuffy cultures (rice, oats), ears, crackers, crispbread, noodles and macaroni. It absolutely injures no safe grains of millet, oil-bearing crops, and bean cultures. Grains with higher humidity are damaged more by the granary weevil than the dry ones.

The weevil is up to 4 mm in length with narrow, cylinder-like, bright body of dark-brown color. Young weevils are brown becoming darker with age. Pronotum is in small oblong pits, elytrons are with deep lengthway grooves. The snout of a male is a bit thicker than in female (Figure 31).

The weevils do not flight and as compared with maize weevil they permeate less through the stored grain. They are very sensitive to draught and grain transfer. Females start oviposition only after some period for additional food. The weevil life span varies depending on temperature and moisture. At room temperature the females live about 14 months, and at 10-12°C for about 28 months; males live for 5-10 months.

The female avoids laying eggs into grains with humidity below 12%. Young weevil does not leave the grain immediately but stays within it for 2-4 days feeding on its floury remains. The granary weevil is a warm-requiring insect. Favorable temperature for its existence is 20-28°C and relative humidity is 75-93%, with the optimum temperature being 25-28°C.

Average developmental time for the granary weevil at 15, 20, 25 and 30°C is 141, 69, 46 and 34 days accordingly. Lower temperature threshold for development is 10.2°C.

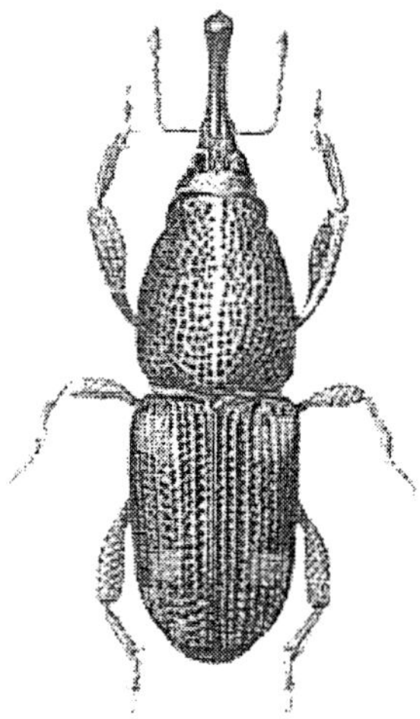

Figure 31. Granary weevil.

Apart from temperature, moisture and food quality, the number of laid eggs and development rate are greatly influenced by the height of the stored grain. Stored grain of 0.5-0.75 m in height, or storage in sacks are much greatly affected by the granary weevils.

In free storehouses and in stored grain with the temperature higher than 16°C the weevil development continues within a year but suspends in winter periods. In unheated premises the development stops at autumn and winter fall of temperature (below 15°C). In unheated premises the granary weevil winters more often as the beetle and in larva stage. They revive after winter dormancy when the temperature in storehouse rises to 8-10°C. The granary weevil beetles can live for a long period of time without food.

In heated premises about five generations can appear within a year, and in unheated premises – from two to four generations depending on geographical location.

Aggregation pheromone [39,40]:

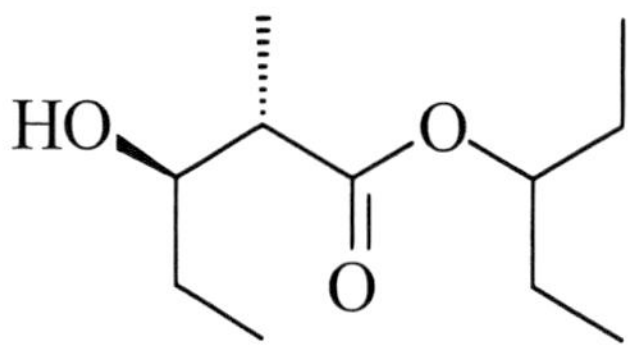

(1-Ethylpropyl)-(2*S*,3*R*)-2-methyl-3-hydroxypentanoat

2.19. Maize Weevil Sitophilus oryzae l.

Maize weevil is widely distributed in the countries with hot and lukewarm climate, and everywhere in heated storehouses; it develops both in the fields (in southern regions), and in warehouses. The weevil and larva damage rice, corn, wheat, barley, sorghum, seeds of pea and beans, buckwheat, acorns, it can also attack flour, semolina, different flour and breadstuffs. Larva feeds on grain content, leaving embryo untouched. Domestic animals do not eat such grain. Maize weevil is more dangerous as compared with granary weevil by the damage made. Maize weevil beetles are 3.5 mm in length. Coloring of their bodies varies from brown to dark-brown, dull. Base and top of elytrons contain lighter reddish spots (by two on each). There are round, shallow and densely located pits (Figure 32).

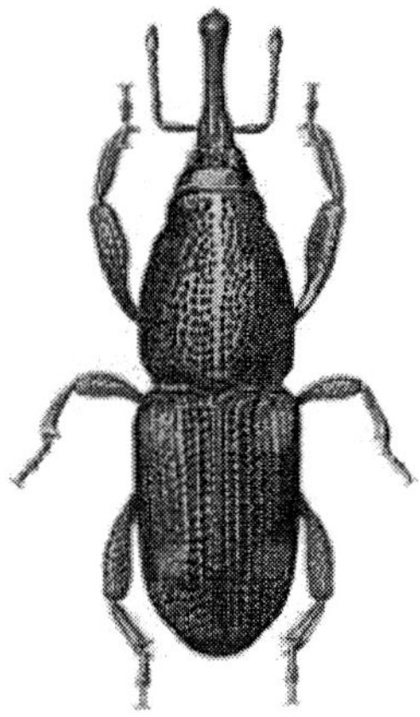

Figure 32. Maize weevil.

Weevil beetles can fly, usually not high, descending rather quick, they are very agile and can cover vast distances; in southern regions they can actively settle from storehouses to fields. Into stored wheat the maize weevils penetrate deeper than the granary weevils. Young beetles need additional food for mating and oviposition. The maize weevil is reproduced like the granary weevil but much faster. At 18°C one generation develops for 3.5 months and at 21-22.5°C for 1.5 months. The weevil can develop inside wheat grains with humidity at least 10%. At optimum temperature (28-30°C) the weevils can live from two to six months. In regions with warm climate and in heated premises they develop during the whole year round, and in regions with moderate climate and in unheated premises they stop reproduction when autumn comes and temperature falls below 15°C. The lower threshold for development is 13°C. There are 7-8 generations per year in heated premises and 3-4 generations in unheated ones.

Aggregation pheromone

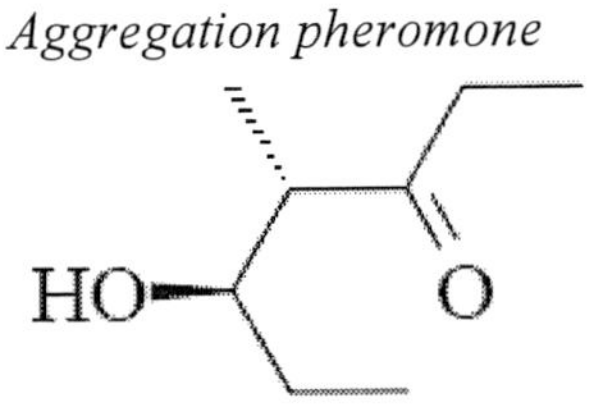

(4*S*,5*R*)-5-Hydroxy-4-methylheptan-3-one

As compared with the granary and the corn weevils, the maize weevil is more sensitive to cold and more resistant to higher temperatures. Eggs are more cold tolerant than pupae and larvae, and the weevils are less cold tolerant. At lack of food the maize weevil dies faster than the granary one.

2.20. Corn Weevil *Sitophilus zeamais mots*

It occurs world wide, in Russia it is found in southern regions. The corn weevil prefers corn, rice and wheat, however, it can also be seen in grains of all other sorts of grain cultures. The damage made is similar to damage made by granary and maize weevils, but weevils eat away all the core of wheat grains, and they enter the corn grains for the second time and feed on its floury portions.

The corn weevil is similar with the maize weevil, and both these species are hard to distinct. The corn weevil is a little larger, its elytrons are brighter with clearly seen saffron spots, however, clear distinction is possible by some signs of male genitals.

Main biological peculiarities including ability to flight are the same as with the maize weevil. The corn weevil is found to have great fertility and it so more resistant to cold than the maize weevil. At zero temperature it lives 18-37 days, and at –10°C – up to 13 days.

Aggregation pheromone:

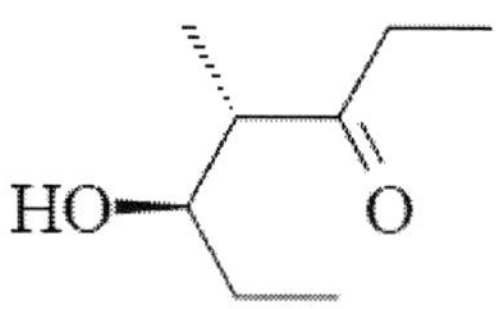

(4*S*,5*R*)-5-Hydroxy-4-methylheptan-3-one

2.21. Angoumois Grain (Grain, European Grain) Moth *Sitotroga cerealella olivier*

Angoumois grain moth occurs world wide except for polar zones. This species is found basically in enclosed spaces (granaries, storehouses, flats), but in southern regions (Krasnodar Territory, the Caucasus and Transcaucasia) it damages directly in fields. It can be seen everywhere in heated premises in Russia, and in nature – in the south of the European part. Caterpillars strongly affect grains of cultivated cereals: corn, wheat, barley, rye, oats, rice, sorghum, etc., as well as buckwheat, soybean, pea, bamboo seeds, but prefer corn. Seeds of legumes are damaged by the moth if their moisture is at least 14%. In granaries the moth is attacking the upper layer of the stored grain (up to 5-8 cm in depth), and at strong infestation up to 20 cm. Besides, before mass flight of butterflies, when the stored grain is strongly infested, temperature of grain is largely increased which results in formation of warming centers. Being within the grain the butterfly feeds on endosperm. Grains left by butterflies differ from grains damaged by other pests by availability of round holes of up to 2 mm in diameter. When opened these grains are characterized by availability of two chambers separated by arachnoid wall.

Wing spread of angoumois grain moth butterflies is 11-19 mm, and the body is 6-9 mm in length. Both pairs of wings are narrow, their length 5 times exceeds their width. Hindwings are with pointed top and hollow before the top. Forewings are of Isabella or sienna-pale color, sometimes with a black spot. Hindwings are silver-grey. Hindwing fringe is wide enough (Figure 33).

Figure 33. Angoumois grain (grain) moth.

The butterflies are active in granaries both in day time and at night, but they fly badly and reluctantly. Imagoes start mating almost immediately after leaving the pupae.

After mating the females can live 13-61 days, and the males – 12-38 days; virgin butterflies live almost twice as much. The females usually lay by one egg on grains (in storehouses) or on ear chaffs in the fields. Parous caterpillars are very mobile and sometimes they are crawling for a long time searching for a suitable place to tunnel into the grain. It is difficult to distinguish grain with the caterpillar inside from undamaged grain, as its inlet is very small and invisible. At grain threshing caterpillars do not die and reach granaries where they still feed on. Pupation usually takes place inside the grain where the caterpillar fed on. Older instar caterpillars and pupae winter.

The life cycle of one generation varies greatly and can last from 25 to 113 days (under optimum conditions 25-35 days on the average). The life cycle at 20, 25 and 30°C is 69, 41 and 29 days on the average [32]. In Russia up to 3-4 generations can be seen in granaries, in the Caucasus – up to 8, among them 1–2 in the field. Caterpillar development greatly depends on the temperature and humidity of grain. At 10–12°C and below the caterpillars diapause which can last for 4–5 months. For normal development the caterpillars need humidity of at least 15–16 %, in dryer grain their development is slowing down, and vast number of caterpillars dies out. At humidity of 8% and less the caterpillars cannot develop.

Sex pheromone [41]:

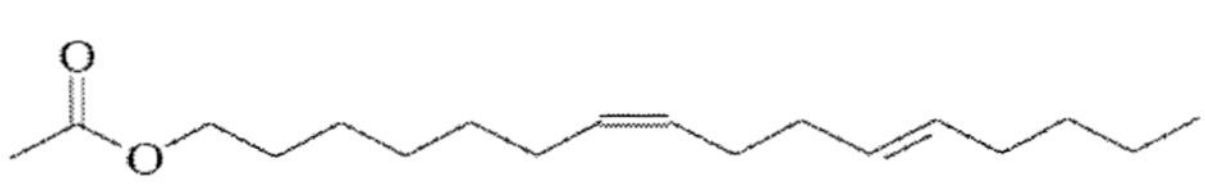

(7*Z*, 11*E*)-7,11-Hexadecadien-1-ilacetate

2.22. Corn Moth *Nemapogon granellis l.*

It occurs throughout the world, Sinanthropus. Wing spread of an adult butterfly is within 10-14 mm (Figure 34). The caterpillar is up to 10 mm in length of yellow color. Eggs are laid openly on food substrate, grain and flour, however, they tend to lay their eggs into cracks of the wall, cracks or folds of packages. Their number can reach 250.

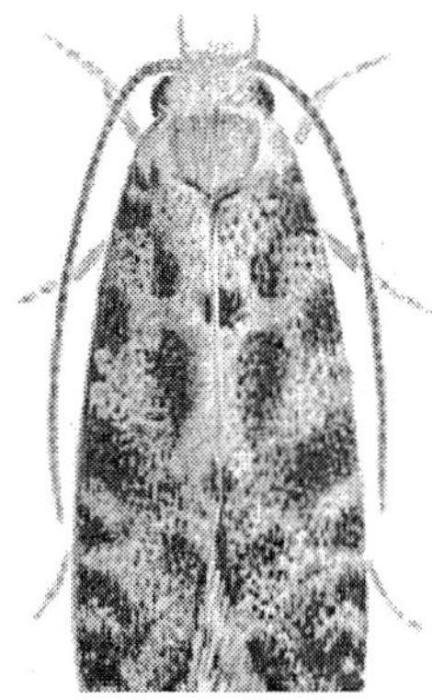

Figure 34. Corn moth.

Eggs are developed in nature within 6-14 days: at 17–18°C development lasts 12–14 days, and at 18–20°C hatching is decreasing to 10 days. Eggs are rather dryness resistant, they do not winter and hatching and development take place in warm seasons only.

When developed on the stored grain the caterpillars make indusia attaching them to the grain and from there they mine in the grain. At feeding it is embryo which is mainly damaged without penetrating inside the grain. During the life cycle the caterpillar crawls over from one grain to another one connecting them by its tubular holes by 2–30 pieces, interweaving excrements and mud, thus forming the so-called nest. The butterflies appear with well formed reproductive products and mating is already possible in several hours. After mating corn moth butterflies live for 5-11 days. In Europe 1-3 populations can develop.

Sex pheromone[42]:

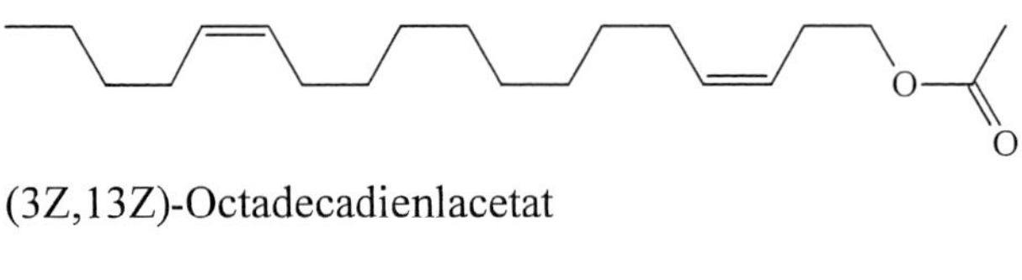

(3Z,13Z)-Octadecadienlacetat

2.23. Clothes Moth *Tineola Biselliella l.*

It occurs throughout the world. The butterflies are about 9 mm in length, wing spread is – 14 mm (Figure 35), they do not feed but lay 100–150 eggs from which in two weeks appear larvae at 25°C. The larvae are of white color with brown head. The larvae life cycle at room temperature is about three months. Up to three generations of moth develop per year in housing premises. The larvae attack wool, fur and other keratin containing materials.

Figure 35. Clothes moth.

Sex pheromones [43, 44, 45]:

(2E, 13Z)-Octadecadienal

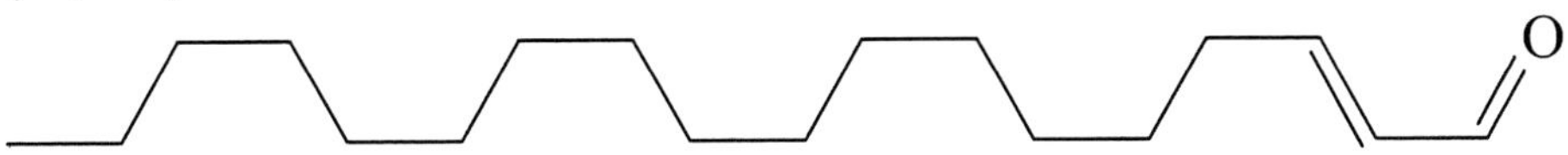

(2E)-Octadecenal

(2E,13Z)-Octadecadiene-1-ole

2.24. Wax Moth *Galleria mellonella l.*

It is found in Europe and Asia. The butterfly has different dimensions (spread from 1.5 to 3.5 cm); forewings are of violet-grey color (in female) or brownish (in males), hindwings are grey (Figure 36). It lays its eggs on honeycomb in behives or storehouses. The caterpillars feed on wax making holes in honeycombs interweaving them with cobweb; honeycombs become unsuitable for bees and for fat clarifying. They can severely injure apiculture making the behives die at mass reproduction. When all the wax is eaten the caterpillars start feeding on excrements left from previous generations, with the new generation of butterflies being far less.

Figure 36. Wax moth.

Attractants [46]:

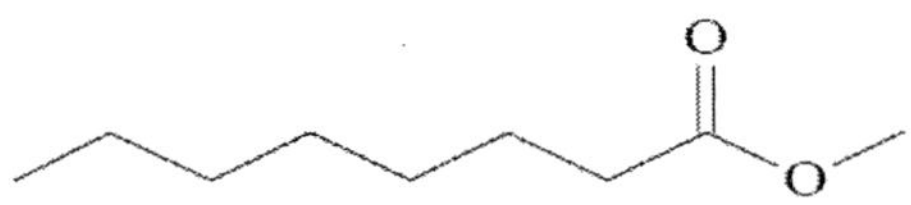

Methyl octanoat

Nonanal

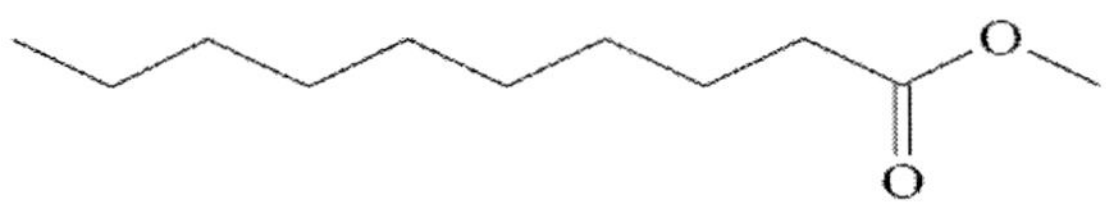

Methyl decanoat

2.25. Rice Moth *Corcyra cephalonica staint*

It occurs world wide, in the tropics it lives in natural conditions. It was taken to Russia together with the imported goods and sporadically found in storehouses and warehouses. Caterpillars damage rice, grain of corn and soyabean, seeds of cotton and other cultures, dried fruit and berries, different products. Wing spread of butterflies of rice moth is 20-25 mm. Forewings are of dun color, almost one-color, front end is a bit lighter (Figure 37).

Figure 37. Rice moth.

Males come out of pupae 2 days earlier than females and live 2-5 days longer. Temperature threshold for development: 17.5-32.5°C, air humidity threshold is 10%. The life cycle of one generation lasts for approx. 30-50 days depending on caterpillar food. Up to 2-4 generations can be seen per year, and up to 7 ones in the tropics.

Sex pheromones [47, 48]:

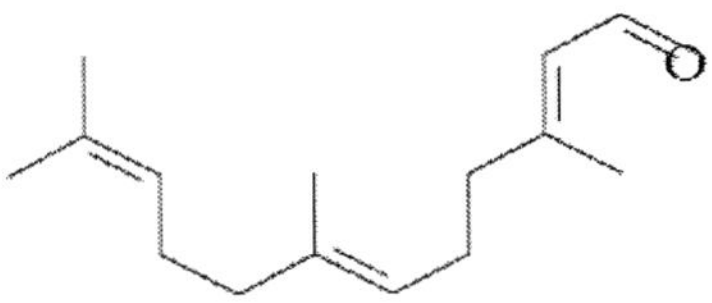

(E,E)-3,7,11-Trimethyl-2,6,10-dodecatrienal

(9Z,12E)-Tetradecadien-1-ole

(6R,10R,14R)-6,10,14-Trimethylpentadecan-2-ole

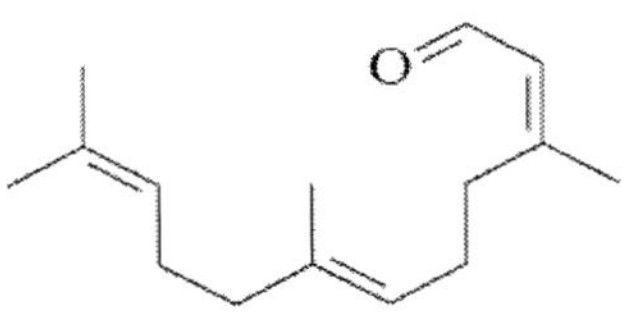

(Z,E)-3,7,11-Trimethyl-2,6,10-dodecatrienal

2.26. Almond (Date Palm) Moth *Cadra cautella wlkr. (= ephestia cautella wlkr.)*

It occurs world wide. In Russia it is found in premises only, more often in southern regions, although it is often taken to other regions with the products. Caterpillars are seen in storehouses, groceries, baggage holds where they affect mainly dried fruits as well as cacao beans, peanuts, almonds, nuts, raw tobacco, grains of cereals and corn, rice, kidney bean, bran, cotton seeds, flour, flour and confectionary goods.

Almond moth butterflies have wing spread of 15-22 mm. Hindwings are wide, their length is almost twice as much as their width. Forewings are of ochre-grey or taupe color, marked with grayish-white spots (Figure 38).

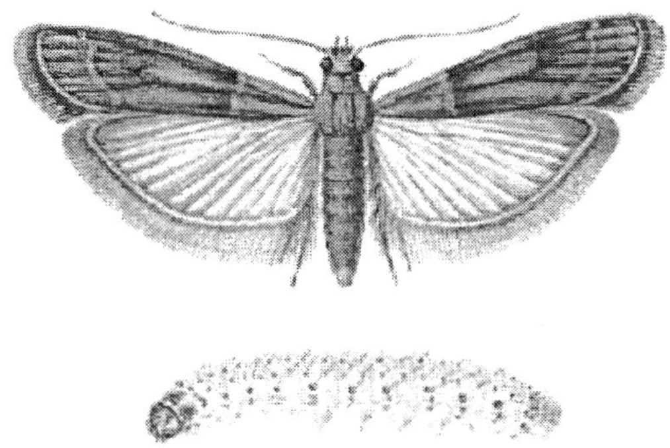

Figure 38. Almond moth.

The butterflies fly in premises since May till October. They develop by the end of the summer and in autumn, and after wintering till April making silky holes and interweaving the nutrient substrate into quite solid clods. Optimum conditions for the Almond moth are 28-32°C, relative humidity about 60%. In moderate climate it can give 1-2 generations, in the south – 2-3.

Sex pheromones [49]:

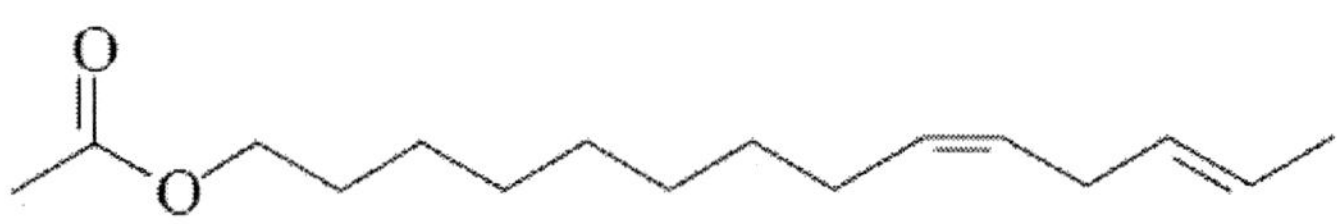

(Z,E)-9,12-Tetradecadienilacetat

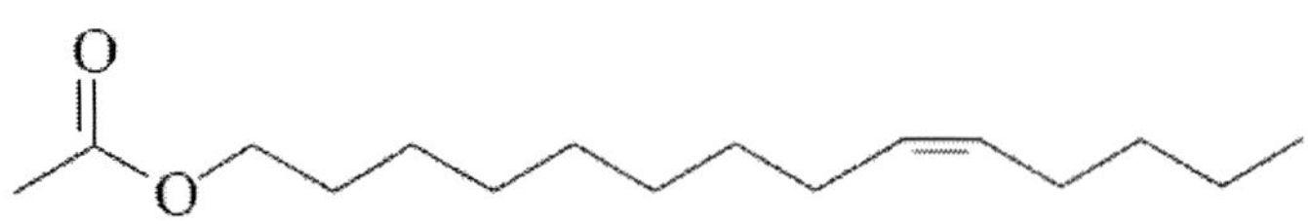

(Z)-9-Tetradecaenilacetat

2.27. Raisin Moth *Ephestia FigulilellaGregs*

It is synanthropic* heat-loving species to be distributed actually everywhere where phytogenous products are stored.

In Russia it is found in premises only, although in southern regions it can develop in nature on mummified fruits, fruit drop, etc. It is also seen in food stores and storehouses, in railway cars and ship holds, in confectionaries. Caterpillars live in tubular yarn on the surface of food substrate and damage different dried and air-dried fruits (fig, raisin, prune, etc.), as well as nuts, cacao beans, cotton seeds, dried herbs, sometimes ground rice and other cereals, including corn. Quite often caterpillars are taken to apartments or flats together with affected

* Synanthropic organisms (from Greek *syn* – together and *antropos* – a human being) – animals which existence is closely related to human beings and human habitation.

goods or confectionery where this species is actively reproduced damaging different products.

Butterfly bodies are about 10 mm in length. Forewings are grayish, hindwings are satiny-white (Figure 39).

Figure 39. Raisin moth.

The butterflies are seen in confectioneries for a period since beginning of March till June and since August till October. Imagoes live for about two weeks. Low development threshold at relative humidity of food substrate of 70% (optimum humidity) is 15°C, the high one is 36°C. Optimum temperature for development is 30°C [50]. The raisin moth gives two generations.

Sex pheromones [51, 52]:

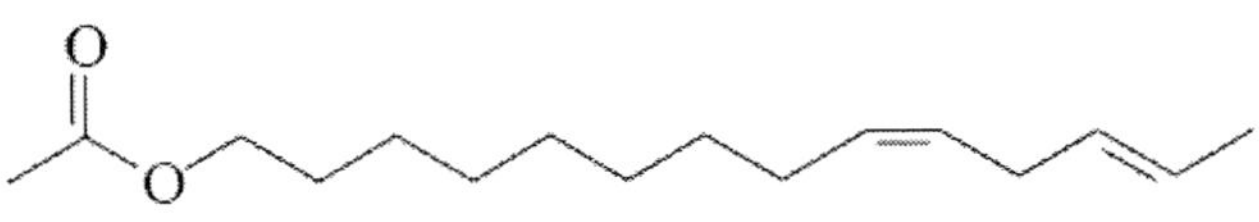

(Z,E)-9,12-Tetradecadienilacetat

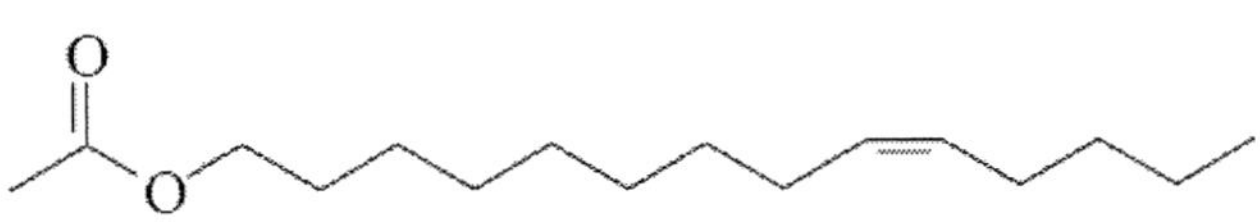

(Z)-9-Tetradecaenilacetat

2.28. Flour (Indian Meal) Moth *Ephestia kuehniella z (= anagasta kuehniella z.)*

Cosmopolite. In Russia it is found everywhere in premises. It inhabits acting mills, bakeries, confectionaries, flour and other food stores, ship holds, etc. In northern and central parts of Russia it can live and reproduced within the whole year in heated premises only. In warm regions in summer time the flour moth is able to reproduce in nature on vegetable remains in landfills, haystacks and threshing-floors.

It damages actually all kinds of plant products: grain (corn including), flour (rye, wheat, barley, corn, soya flour), all sorts of cereals (esp. semolina and buckwheat), milled rice, bran, macaroni, crackers, confectioneries, chocolates and soya chocolates, nuts, dried fruits, vegetables and mushrooms, cotton seeds. Sometimes it develops on kitchen wastes, it can injure insect collections. More often caterpillars feed on broken grains rich in vitamins, proteins and fats: embryos and glumes. Among products they prefer grains of corn and soft wheats of medium and high humidity, therewith damaging endosperm as well.

Wing spread of flour moth butterfly is 17-27 mm and the body of 10-14 mm in length. Forewings are dark or ash-grey. Wing pattern consists of two lighter indented diametral lines bordered by blackish edges. Hindwings are wide, their length is at least two times more than its width (Figure 40).

Butterflies appearing from pupae do not scare of scattered light, they often fly in day time, however, in premises they are more active in the evening. In day time they usually can be seen in more or less shadowy places of walls, ceiling, partitions, or inside machines. The butterflies appear with well formed reproductive products and mating is already possible in several hours. The majority of species mates within 1-3 days. After oviposition proceeding for 5-14 days, the female lives 3-4 days more. Virgin females live longer up to 20-25 and more days, however their ability for mating and fertility decrease significantly. Low temperature slows down the oviposition process, decreasing real fertility and somewhat increasing life time.

Females start laying eggs in several hours after mating. The eggs are placed by one or rarely by dozens into cracks in walls, grain bins, on containers (sacks, canvas), dusty parts of equipment, into gaps of gravity conduits, etc. Caterpillars appearing from eggs release large amount of cocoon silk and make tubular holes penetrating into the upper layer of flour, and in stored grain they bore into 10-15 cm.

Caterpillar development is greatly influenced by the character of food substrate and temperature. Feeding is delayed on products containing small amount of proteic substances. Low temperature in autumn-winter season is suspending development for up to 170 days, and at 8°C it stops. After finishing feeding caterpillars of the last, 5th instar, are looking for a place for pupation. At that they are leaving food substrate and crawl into cracks in walls, grain bins, pillars, partitions, sack folds, etc., where they net fusiform silk cocoons.

Figure 40. Flour moth.

Depending on food, temperature and moisture the whole life cycle lasts 36-270 days. Thus, at optimum temperature of 25.8-27.8°C it is finished for 36-42 days, in summer – 50-65 days, if temperature falls below 18°C life cycle lasts for 3 months. In autumn and winter if temperature varies within 8-15°C the development cycle lasts 157-270 days. There are data of availability of two races of pests: with fast type of development and slower development under similar environmental conditions. Development rate of "fast" race is almost one and half time higher than in "slow" race.

In moderate climate of Eurasia the butterflies more often can be seen in unheated premises of granaries and grain processing plants for a period from April to May and from July to October. In heated premises they are found the whole year round. Number of generations per year is different. In unheated premises of temperate zone it gives 2-4 generations, in the south – up to 6 generations. Under favorable climatic conditions up to 8-10 generations per year are likely to appear in heated premises.

Attractants [53, 54, 55]

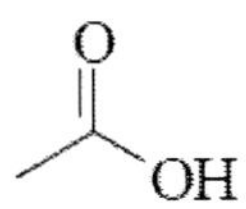

Acetic acid

Kairomones [56]:

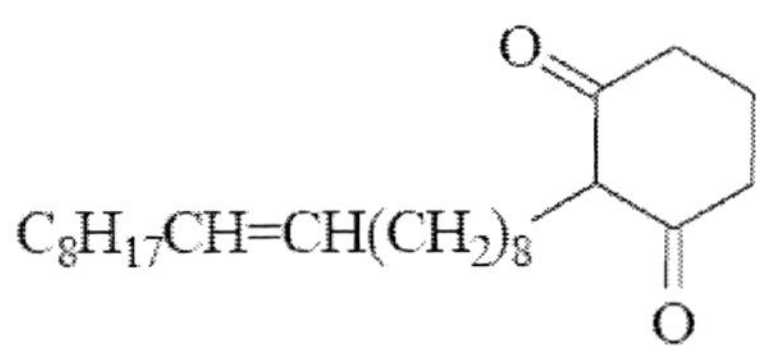

2-(9'Z'-Octadecenyl)-cyclohexan-1,3-dion

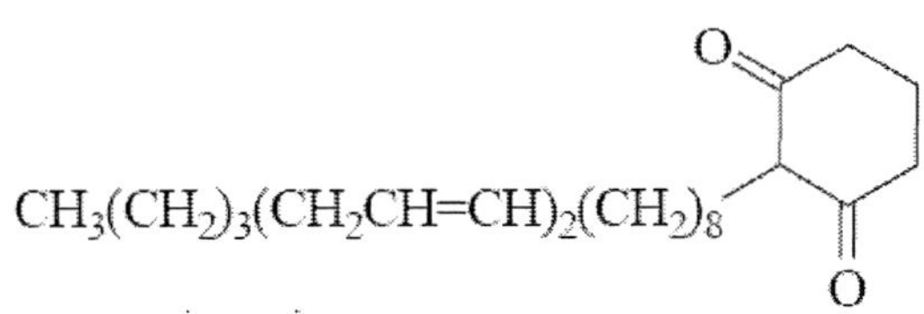

2-(9'Z,12'Z)-Octadecadienyl-cyclohexan-1,3-dion

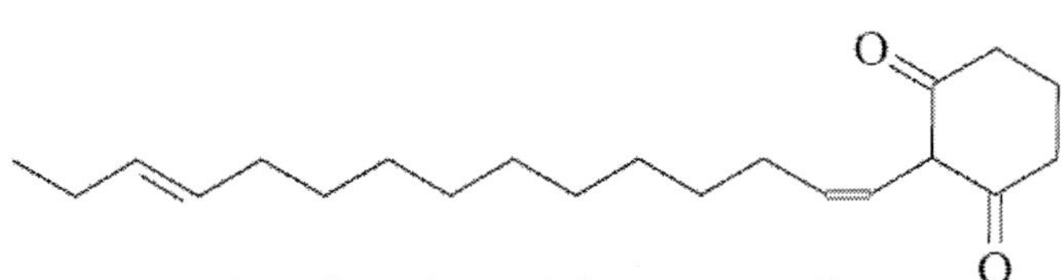

2-(12'Z,E14'-Hexadecadienyl)-cyclohexan-1,3-dion

Sex pheromones [57]:

(9*Z*,12*E*)-9,12-Tetradecadienale

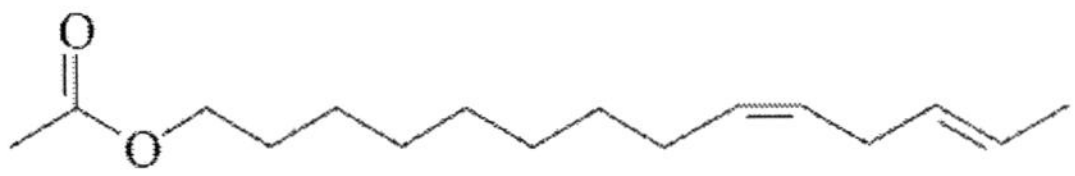

Z,E-9,12-Tetradecadienilacetat

2.29. Cacao (Warehouse, Tobacco) Moth *Ephestia elutella hb*

It is actually found everywhere in premises, it develops out-of-doors in the tropics only, in temperate zone it is rarely seen, it is due to mass flight of butterflies only. In Russia in unheated premises it usually occurs in the middle of the country and in the south of European part, in the northern Caucasus and in the south of Siberia; in heated premises it can be seen e.g. in confectionaries, and quite often in other regions.

It is quite often found in warehouses, storehouses, ship holds, in confectionaries, sometimes in apartments. It damages all sorts of grain (wheat, rye, corn, rice, different cereals), bakery products, confectionery (flour, crackers, pastry, spice-cakes, biscuits, candies, chocolates), cacao-beans, coffee-beans, soybeans, dried fruits, peanuts, almonds, dried vegetables, hulled sunflower seeds, mill cakes, bran, crude drug, tobacco and tobacco goods, seeds of garden plants and of forage herbs, as well as herbaria and collection of dried insects. It is usually seen on dried fruits and confectionary. It is one of the most serious pests of long-term stored grain.

Cacao moth butterflies have wing spread of 12-20 mm and the body length is 6-8 mm. Forewings are ash-grey with two light ties and dark edgings. Hindwings are wide, with their length being twice as much as their width (Figure 41).

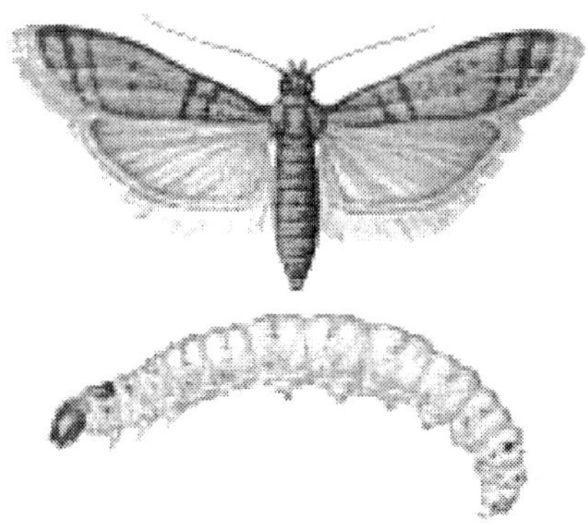

Figure 41. Cacao (warehouse, tobacco) moth.

Imagoes do not need additional food. Mating can take place at any time of the day. The majority of escaping butterflies mate within the next 1-2 days. Mating continues for about one hour. Imago life cycle in the laboratory conditions is up to 9 days in males and up to 17 days in females; and in natural conditions it limits up to 10-14 days. Butterfly can be active at any time of the day, and flight to light sources is poorly expressed. Correlation between males and females in different seasons is different, but, as a rule, females always prevailed.

Caterpillars leaving eggs crawl fast and soon they tunnel in glume feeding on its embryo. After 1st or 2nd molting they leave the first grain and start eating another one but from inside. For the whole life cycle the caterpillars damage embryos of 33-49 grains each. Caterpillar developmental time is greatly influenced by substrate feeding power and temperature and sometimes it can be much expanded. In laboratory conditions at relative humidity of 70% the caterpillar development on wheat lasts 54-80 days at 17°C, and at 25°C (optimum temperature) for 29-50 days only. Lower temperature threshold for development is about 10.4°C. As a rule, older instars caterpillars pass the winter. Wintering takes place in narrow cracks of wooden walls of granaries, rarely within balls of grain on the surface of stored grain. The caterpillars therewith swarm in large communities, although each one is located in a separate light cocoon.

In moderate climate it gives 2-4 generations per year in storehouses. In European part of Russia second generation is often seen which significantly exceeds the first one by abundance of flying butterflies, as the majority of butterflies stay for wintering. The whole life cycle in summer under optimum conditions lasts for 44-90 days, but it is expanding for several months if the caterpillars diapause or winter.

Sex pheromones [58]:

(Z,E)-9,12-Tetradecadienale

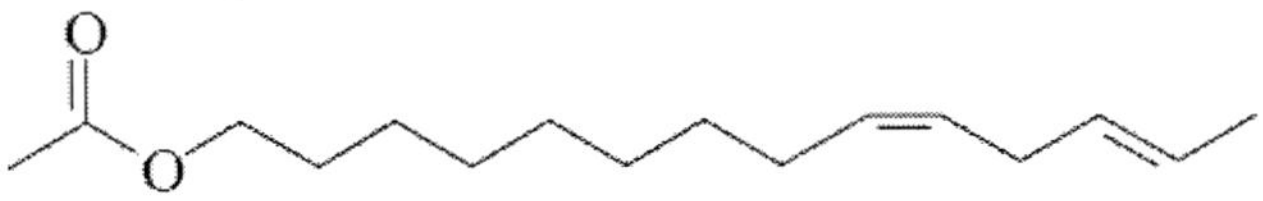

(Z,E)-9,12-Tetradecadienilacetat

2.30. Indian Meal Moth *Plodia interpunctella hb*

Cosmopolite. In Russia in warm premises it occurs actually everywhere, except for polar regions of Siberia and the Far East.

In temperate climate this heat-loving species develops almost exclusively in premises, but in the south of European part of Russia and in the Caucasus it can give in summer 1-2 generations in nature.

Indian meal moth is usually seen in warehouses, back rooms of groceries, in flour mills, bakeries and breweries, confectionaries, sometimes it is found in apartments. It is one of the most serious pests of stored products. It feeds principally on dried fruits and vegetables, herbs, grain, cereals and products of their processing, as well as on cocoa beans, different legumes, kernels of walnuts, sunflower seeds, confectionery (esp. candies), spiceries and many other phytogenous products. Sometimes caterpillars damage herbaria, collections of dried insects, old honeycombs and beebread. They are also seen in nest litter of different birds. There are cases of cannibalism when actively feeding caterpillars eat pronymphes. In southern regions they severely injure dried fruits in canneries and stored grain, being even more dangerous than meal snout moths.

Indian meal moth butterflies have wing spread of 13-20 mm, and their bodies of 7-9 mm in length. One third of a forewing is whitish-yellow, two thirds are of russet color (Figure 42).

Imagoes do not need additional feeding. The butterflies fly in the evening and at night, and in day time they hide in shadowy places of warehouses. Males are more agile than females. The majority of species evolving from pupae mates within the first 1-2 days. Imago life cycle depends both on mating, and on temperature, and at 19.4°C it lasts for 12-19 days, and at 30°C – 5-6 days only. As a rule, males live longer than females. Butterflies can be seen in the ceiling, walls, equipment, and products in premises affected by the moths.

Parturiated butterflies crawl fast looking for cracks or just cavities in food substrate. On the 2nd or 3rd day they start making tubular holes covered with excrements from which they feed. Sometimes, when developed on dried fruits they do not make tubules and caterpillars bore into the fruits. In summer time in stored grain caterpillars live in upper layer interwining it with silk thread for up to 10 cm deep, however, at cold temperatures they tunnel in deeper and in January-February they can be found 1.5-2 m deep from the surface. Depending on the season caterpillar developmental time varies from 28 to 56 days, and under unfavorable conditions it can be 267 days. Food quality also affects it.

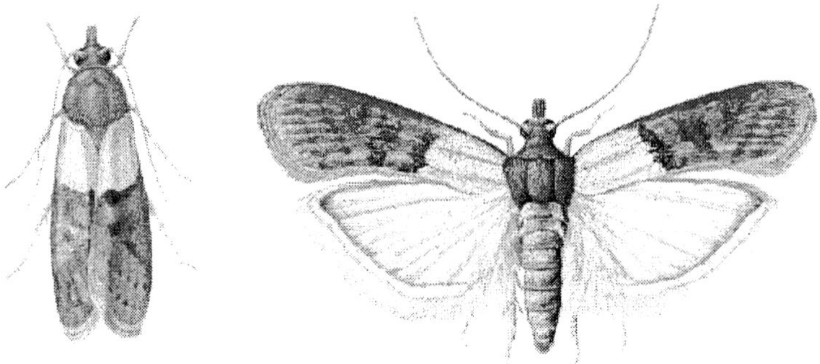

Figure 42. Indian meal moth.

The whole life cycle of the Indian meal moth lasts in summer for 60-85 days, which, depending on the ambient temperature and suitable food for caterpillars, results in 1-6 generations yearly. In laboratory conditions at constant temperature of 21.1°C developmental cycle of one generation was decreasing to 42-56 days, an at 27-31.1°C – to 36 days. Lower temperatures expand life cycle for 9-11 months. Optimum temperature for development is 28-

32°C. Lower temperature threshold for development is 14.3°C. At 20, 25 and 30°C one generation develops for 62, 33 and 23 days accordingly on the average.

Pheromones [59]:

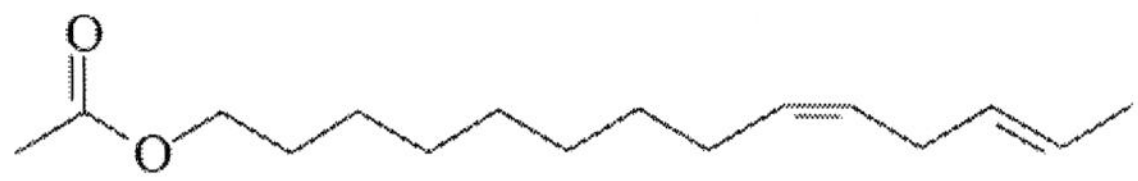

(Z,E)-9,12-tetradecadienilacetat

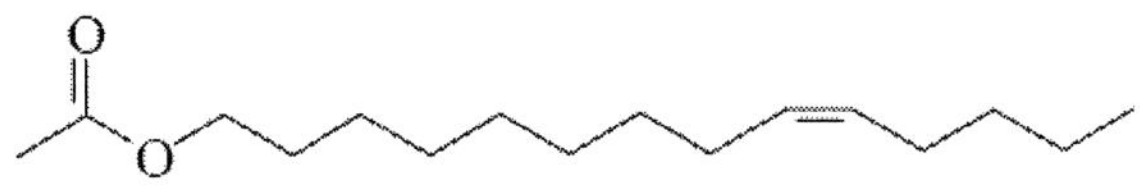

(Z)-9-Tetradecenilacetat

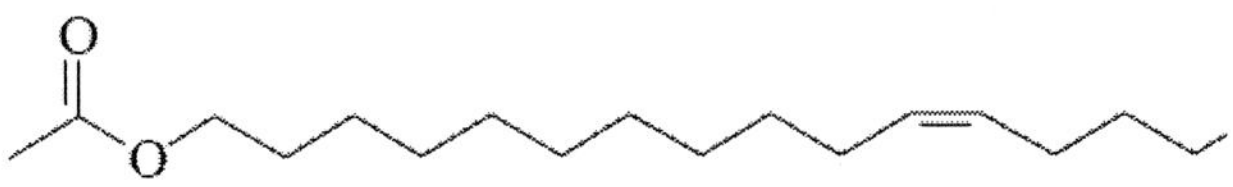

(Z)-11-Hexadecenilacetat

Attractants

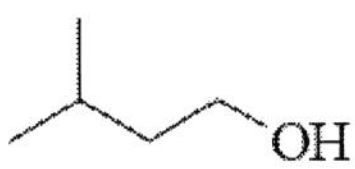

3-Methyl-butanole-1

Chapter 3

3. Rodents

Until recently these mammals were considered to be serious vermins of both foodstuffs and nonfoods. They all belong to rodent order – *Rodentia,* to one family – murine (*Muridae).* The name of the order is due to having constantly growing intermediate incisors for the account of constant embryonal tissue in teeth root and self-sharpening due to friction of upper and lower incisors on each others.

The rodents can bee seen world wide. Their biological peculiarity is ability for quick reproduction which is specified by large number of young in litter, multiplicity of litter due to all year round reproductive activity as well as due to early sexual maturity of litter. Many rodents feed on plant matter, but those species which accompany man are actually omnivorous. High ability for reproduction results in rodent plagues, which was called at all times as "mice attack". There are only three species which chiefly damage the stored goods: the house mouse *Mus musculus*, the field mouse *Apodemius agrarius* and the common rat *Rattus norvegicus*. Quite often the black rat can be found among the imported goods.

The house mouse *Mus musculus* is a small creature. Its body is 7-10 cm long, with its tail (covered with annular arranged horny scales and rare hair) being 50-100% of the body length. Fur coloring of desert forms is light, yellow-sandy with absolutely white belly, and coloring of northern forms is generally known "mousy-gray" on the back and sides and light-gray on the belly (Figure 43). They hide in burrows which are not large and simply made: with nest cell located 20-30 cm in depth and usually with one exit. But they prefer to settle in burrows of other rodents: Microtus transcaspicus, mole-vole, gnawer beetles, etc. Usually they occupy free or unattended parts of inhabited burrows. House mice settle in close proximity to humans as well, but they are not specially attached to them. At any time of the year they can settle in dwellings and leave them. The mice reproduce in desert zones during all the warm period from March till November. They give birth to 2-3 litters, from 2-3 to 9-10 (usually 5-6) young in each. In heated premises they reproduce in winter period as well.

The field mouse *Apodemus agrarius* is larger than the house mouse. Their body is 10-12 cm long, tail – 6-9 cm. Fur is tawny from above and on the sides (the young have soft guard hair, the adults have coarse one). There is a clearly seen black stripe ("belt") from nape to tail bottom along the middle of the back. The field mouse is distinguished from other mice species of Russia fauna by this stripe. They rarely settle in buildings and in small amounts, and quite often are seen in winter-crop and row-crop plantings, avoiding softwood forests. They hide in burrows dug up by themselves or other rodents and moles.

Figure 43. House mouse Mus musculus.

The mice feed on seeds and green parts of plants, sometimes adding some insects to food. In warm season the mice give birth up to 3 litters by 3-9 young in each. The field mice are supposed to transfer hemorrhagic nephrosonephritis, Q fever (from rickettsiosis) and one of leptospirosis forms.

Rats *Rattus*, rodent class (mouse family - *Muridae*). Its numerous species are spread throughout the world. Many other average size rodents are often called as rats, adding some adjective to them (e.g., kangaroo rats, large-toothed rats, rock mice, earless water rats, big-eared rats, etc.), however, generally word "rat" means mainly the common rat or the sewer rat *Rattus norvegicus*. This generally known species is found in majority of temperate zone cities, especially in Europe, in the north of the USA and in Canada. Albino rat (some sort of a red one) is bred in laboratories and is kept as a pet. The black rat (*R. rattus*) lives next to people throughout the Mediterranean, in the south of the USA, Central and South America. The rats' habitat was initially the Old World but now they have inhabited all continents and most islands.

Over 550 taxa of rats are known. The body of various species is from 13 to 30 cm long, excluding a squamose tail of about the same length or longer. Average size common rat weights appprox. 0.7 kg, its body tail excluding being 23 cm long (Figure 44). Rats breed almost all year round and their population can grow rapidly. The pregnancy usually lasts 21 days, the number of the young in a litter is from 3 to 12, average 8. Little rats are born naked and blind but they grow fast. They open their eyes on the 15th day and in 3 weeks they can look for feedstuff themselves. At the age of two a rat stops breeding and by the age of three it grows decrepit and loses its teeth.

The rats are social animals, but there are often fighting between them. The most vulnerable part of their bodies is the tail; scars from wounds made by other congeners are often seen on it. Birds of prey (e.g. eagle owls) and animals (martens, foxes, etc) are chasing them, however, as they live in proximity with humans these predators are not usually threatening them. As far as the cats are concerned they hardly dare hunting for such a large rodent.

Rats can spread typhus, bubonic plague, icterus, tularaemia, rabbies and trichinosis. They are primarily seen in landfills and sewers, and carry infections by spoiling stored foodstuffs.

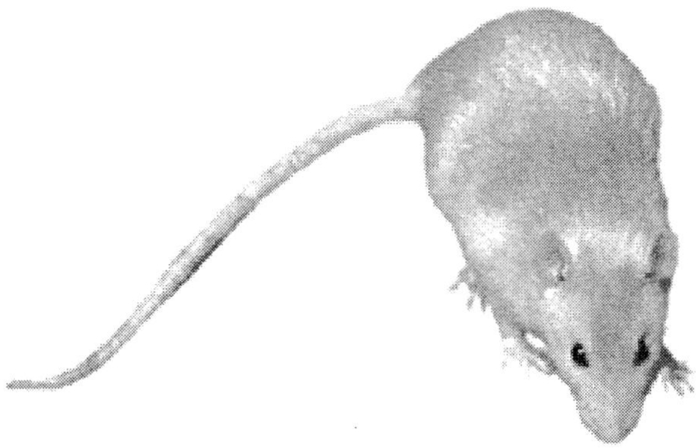

Figure 44. Rat.

The rodents cause damage not only by eating stored products, but by gnawing many things uneatable for them Rodent control is made by using mechanical means (baits) and chemical preparations (rodenticides). The complexity for applying chemical preparations is that poisons toxic for mice and rats are also toxic for humans, and the only protection for the humans against poisoning is large mass of a body.

Earlier for rodent control virulent toxicants (e.g. lead phosphid) and anticoagulants were used, at present for this purpose furocoumarins and biologicals are used which result in epizooty among rodents – suddenly and temporarily affecting a large number of animals.

Chapter 4

4. Methods for Biopolymers Pest Control

Preventive measures for protecting stored goods include thorough selection of packages for them. Hermetically sealed containers are not possible in all cases, however recent technologies offer the most suitable ways for packaging depending on stored goods and materials.

One of the most important means in thepest control system is thorough inspection of imported goods. The primary quarantine examination is made in border crossing points and in places where the goods are delivered. The secondary examination to be made in destination points: warehouses of import raw material processing plants and in storehouses of commercial organization, is also very important. Insects in different stages of development discovered among delivered goods, molting skins (exuviae) of caterpillars and pupae, cocoons, cobweb, excrements and damages are signs of pest availability.

As a rule, the most efficient way is combination of different methods, however, they are not always possible to apply.

Physical methods include insulating layers, mechanical impact (cleaning), as well as high or low temperatures, radiation treatment and electric current effect.

Mechanical cleaning is efficient if the pest is not quick enough. It helps in moth and dermestids control, but it gives no effect with cockroaches or ants.

Heat treatment can be made by applying low temperatures in warehouses which is quite easy to be made in winter seasons, and to safely protect against actually all storage pests. Besides, to prevent development of insect eggs, larvae, pupae developing directly within stored fabrics, leather, furs or finished goods they apply high temperature – pressing, where it is possible, or heating in the sun or by premises ventilation with hot (60 - 70°C) air.

Ultraviolet irradiation is made by applying quartz lamps in premises which simultaneously kill the pests and destroy harmful microorganisms, or it is substituted by solarization.

Exposure to high-frequency currents is short (8-10 min.) but efficient method for control of pests damaging goods. Along with it, uniform heating of goods takes place but it requires special equipment, i.e. electromagnetic plates for making high-frequency field and conveyor for treating large amounts of fabrics, materials or goods.

Biological methods mean rejection of preparations with strong poisons being their active substance, obtained synthetically and application of *living organisms* or their metabolites in pest control. Biological methods include application of entomophages i.e. predators, parasites (mostly nematodes) and disease microorganisms i.e. protozoa, fungi, bacteria, viruses, or preparations based on their toxins.

In some cases *genetic methods* are applied, e.g. artificial introduction of sterile males of predator species which results in maximum decrease of breeding power and extinction of pest population.

Biological pest control is being less developed than other methods due to peculiarities of protected objects. Introduction of predators and parasites into stored foodstuffs is impossible as it spoils products and their quality. Application of microorganisms is the most effective. At present, numbers of preparations based upon different strains of entomopathogenic bacteria and other pathogenic microorganisms were elaborated to be applied for protection of stored foodstuffs and nonfoods.

Chemical methods are still the most applicable and effective ones in pest control. Substances used for chemical treatment are divided by impact into contact or digestive poisons and fumigants acting like gases.

Contact poisons are those that have the ability to penetrate the intact body surface. They are applied as liquid preparations, or as powders i.e. dusts.

Digestive poisons are applied as liquids sprayed onto treated surface, and as powders or grains often together with baits, and as gels.

Fumigants are toxic agents in gaseous form. They are effective to be applied in closed premises as they penetrate all the surfaces to be treated. They can be used as sprays or gases, as well as in preparatory form of slowly evaporating active substances i.e. grains, shells, pills, and plates.

By spectrum of activity chemical insecticides applied now for protection include poisons with specific impact upon insects which is possible due to significant differences in metabolism of both insects and humans. Such substances include known long ago organochloride compounds (DDT), organophosphorous compounds and pyrethroids i.e. fast-acting synthetic analogues of natural substance (pyrethrum), affecting insect nervous system, derived initially from Dalmatian pyrethrum.

All applied chemical insecticides are specially affecting the designated part within the insect body (Table 4).

Neurotoxicants have one common feature, they extensively stimulate nervous system. It its turn, it can result in irreversible changes in metabolism. External symptoms of neurotoxicants action are overactivity (activation, motion hyper activity), tremor (extremities shudder and incoordination of movements), fits and paralysis.

*Organochlorine compounds*They are world famous insecticides: DDT and a number of its analogues with chlorine atom in the active center. They are able to block ion transport mechanism within nerve cells, thereby quickly destroying normal functions of their membranes and suppressing insect nervous system on the whole. As nervous system coordinates with all vital processes, its dysfunction results in quick death.

Table 4. Insecticide classification

Group name	Trade marks examples	Mechanism of action
Organochlorines 1	DDT	Reversible activation of sodium channels of neurocyte membranes
Organochlorines 2	Hexachlorocyclohexane. Lindane. Polychlorcyclodienes	Inhibition of inhibitory synapses of the central nervous system (CNS), mediator-4-aminobutyric acid
Organophosphates	Dichlorvos Chlorpyrifos Diazinon Fenitrothion Azamethiophos Pirimiphos-methyl Malathion (karbofos) Trichlorfon (metrifonate), fenthion (sulfidofos), temephos (diphos)	Irreversible inhibition (phosphorylation) of cholinesterases in CNS excitatory synapses
Carbamic acid derivatives, carbamates	Sevin (carbaryl) Propoxur Bendiocarb	Analogs of acetylcholine – mediator of CNS excitatory synapses; reversible inhibition of acetylcholinesterase
Pyrethrins and first generation synthetic pyrethroids	Permethrin Ethofenprox Pyrethrins, allethrin, tetrametrin	Reversible activation of sodium channels of neurocyte membranes; reversible depolarization of cell membrane and nerve conduction block
Pyrethroids cyano- and fluorine-containing	Biphenthrin Lambda-cyhalothrin Alfa-Cypermethrin Esfenvalerat Zeta-Cypermethrin Deltamethrin Cyflutrin Cypermethrin Cifenotrin Beta-cyflutrin	Irreversible activation of sodium channels of neurocyte membranes; irreversible depolarization of cell membrane and nerve conduction block

Table 4. (Continued)

Group name	Trade marks examples	Mechanism of action
Sulfonamides	Sulfluoramid	Suppression of cell respiration and protein synthesis disorder
Aminohydrazones (hydrazones)	Hydramstillyon	Suppression of cell respiration, slow neuron necrosis
Insect growth and development regulators	Fenoxicarb Hydroprene, methoprene	Analogs or antagonists of juvenile hormone; analogs of molting hormone. Development program disorder
	Difluorobenzuron, Triflumuron	Chitin synthesis inhibitors – disorder in transporting chitin precursor
Avermectines	Avermectine, abamectine	Receptors blocking of γ- glutamic acid in myoneural junctions
Toxin *Bacillus thur-ingiensis*	Bitoxybacillin, dipel, lepidocide	Disorder of cell respiration and ATP synthesis; disintegration of intestinal wall cells
Nereistoxins	Nereistoxin, tiocyclam, bankol	Cholinergic synapses blocking
Cloronicotinoids	Imidaclopride, acetamipride	Irreversible blocking of postsynaptic nicotinic acetylcholine receptors in CNS excitatory synapses
Pyrroles	Chlorphenapyr	Disorder in breathing processes of mitochondria
Phenyl pyrasols	Fipronil	Blocking of GABA receptors in CNS inhibitory synapses

Organophosphorous compounds (OPhC) also belong to neurotoxicants. They differ from organochlorine toxins not only that in active center there is phosphorus, but in fine mechanism of action. OPhC suppress acetylcholinesterase* (AChE) activity – enzyme responsible for impulse transmission from one nerve cell to another one. Like other neurotoxicants, OPhC very quickly destroy insect normal activity, paralyzing their nervous system.

Like the organophosphates in their activities *carbamates* are selective AChE inhibitors. Unlike OPhC the AChE activity suppression has reversible character. Such insecticides like

* Acetylcholinesterase – an enzyme in nerve cells that is responsible for the destruction of and thus for switching off excitation of the nerve.

nicotinoids and nereistoxins also destroy impulse transmission, however, their aim is not AChE but acetylcholine receptor. These toxins block these receptors irreversibly contacting them.

At present the following pesticides are applied. Table 4 gives information on them and on mechanisms of their toxic action on insects.

Hormonal action substances make a special group. They are *analogues of juvenile hormones* (AJH), preventing normal insect ontogeny as well as disturbing reproductive processes – maturation of ova and eggs development. AJH and analogues of ecdysone – *ecdysoids* are also called hormone-mimetics. Preparations of this class are not dangerous for humans as the warm-blooded have no similar processes in metabolism.

AJH do not act like poisons in the common sense. Their action is related to the fact that during larval and pupal development within the insect body the normal development program starts, switches over and is controlled by titer modification (concentration) of natural hormones in the body, precisely timed to periods of molts, pupation, imago appearance, formation of reproductive system, and diapause states. Critical period (when titer of JH or of ecdysteroids is sharply decreased) is the point when hormone-mimetics application disturbs normal ontogenetic processes. The result is additional larval ages, appearance of intermediate forms – larval-pupal, pupal-imaginal, and semimetamorphosis in insects – larval-imaginal. All these species die in the process of development. Besides, AJH possess large number of delayed effects: suppression of reproductive system resulting in either aplasia of this system in treated generation, or in depression of vital activity of species to be appeared anyway in future generation. Diapause interruption also leads to decrease of insect survival rate.

Research of biological activity of hormone-like materials starts with their extraction from endocrine organs of insects. Application of methods of preparative liquid chromatography made it possible to elaborate methods for releasing natural juvenile hormones, as well as important molting hormones – ecdysones from adjacent and cardiac bodies in majority of insect species. Spectroscopic analysis methods made it possible to identify components and to define main reactants of insect hormones. Thus, it was discovered that natural juvenile hormones found in insects in different combinations have isoprenoid structure and ecdysones regulating molting process have steroid structure. Later on it was obtained great number of biologically active analogues of insect juvenile hormones both extracted from vegetable substances and synthesized. Several classes of organic substances appeared to possess juvenoid activity, with their activity being evaluated through a number of bioassays in standard laboratories (pupae of yellow mealworm beetle, flour moth, larvae of elder instars of some insect species).

Chitin synthesis inhibitors (CSIs) are also an insecticide group specific for insects and a number of other arthropoda having chitinized covers. Basically these substances are intestine-affecting disturbing chitin forming process within the insect body at the stage of transporting monomers-precursors (N-acetyl-D- glycosamine) to places of polymer chitin synthesis. CSIs are very effective against insect larvae, they interrupt normal formation of new covers at molting.

Antifeedants are substances preventing substrates treated by them from eating. They present a group of *repellents* – compounds not killing but scaring insects. Among these substances there are many natural compounds contained in plants.

Pheromones and synthetic analogues of insect natural exohormones are the means for chemical communication between one and the same specific species. Sex pheromones especially effective in combination with glue baits are often applied for pest control.

Lepidoptera pheromones are volatile compounds presented by long polyolefin and diolefin molecules.

Insect pheromone composition is analyzed by gas chromatography and chromatography-mass spectrometry, working therewith with extracts (extractions) of insect pheromone glands.

Identification of main pheromone components is followed by selection of synthetic compounds homologous to extracted ones, and evaluation of their physiological activity. To make this evaluation the laboratories use such devices as olfactometers and aerodynamic tunnels for registration of insect behaviour reacting to the tested compounds. Sometimes synthetic analogues are more efficient than the natural ones. Mixture activity of identified compounds is to be evaluated, because it helps to get synergistic result – multiple potentiation of active substance by adding small amounts of other components.

Development of efficient and economic technologies for synthesizing pheromones and making preparative forms for their practical application is very important for all chemical means for protecting stored goods.

Chapter 5

5. Methods of Pesticide Application

Pesticide effectiveness greatly depends on the methods of their application. The simplest way was using of powders obtained by reducing parts of plants or mineral substances. However, solutions or extracts of different origin substances proved to be more effective in number of cases.

For the last years world scientists make research in elaboration of new preparative forms to significantly increase efficiency of used substances. One of such forms is e.g. encapsulation, making it possible to dose exudation of acting substance into environment and avoid undesirable contact of humans and animals with pesticides.

Sprayers can contain mixture of several hermetically sealed insecticides. The majority of them have small amount of active substance and are intended principally for treating small areas. Until recently sprayers solutions have been mainly used for killing insects in living premises. At present the sprayers are used in the open air after making powerful spraying plants.

Water-soluble concentrates are usually liquid compositions containing 18-75% of active ingredient, surfactant and antifreezing agents preventing from freezing at low temperatures.

Granules are dry compositions consisting of inert granules treated with active ingredients of 30-60 mesh* size, such as clay, vermiculite or sand. They usually contain 1-20% of insecticide. These compositions are applied immediately without preliminary mixing. The granules are easy to be transported, storing, discharging and dosing, they possess antistaining coatings and odorless, and are not toxic for the skin. The granules are soluble in water for up to 50 mesh, do not plug usual agricultural sprayers when applying in the field conditions. The granules are heavy, and it results in insecticide dispersion minimization and contamination of areas not to be treated anyway. The granules are rather used for treating soils than leaves.

Dispersed or water-soluble granules are dry mixtures. Ingredients are in granules easily dispersed in water. To avoid aliquation, the container with the solution to be dispersed is to be agitated.

* *Mesh* is a way for measuring powder size fractions. It is measured by passing of particles through a sieve with a certain number of threads along the one running inch (1 inch = 25.4 mm). So, grains of 100 *mesh* size are a powder passing through a sieve with 100 openings per inch, but not through the next sieve with larger number of openings. The smaller are the openings of the sieve, the thinner is the powder.

Dusts are dry mixtures of insecticides with inert powders such as organic substances, minerals, talc or clay. Dusts consist of thin particles of approximately 250-350 mesh. They are usually prepared with concentration of active ingredients of 0.5-10% and are applied without additional preparation. Dusts can be applied to actually any surfaces without damaging them, but visible powder usually spoils environment. Due to small particles the dusts are easily blown out after application. They are effective when dry, but after becoming damp they become compressed and ineffective. However, low moisture of plants can actually help in distribution and adhesion of dust to surfaces. Dusts are of little use in protecting large areas.

Mineral oil emulsion is a liquid composition containing active ingredient solute in refined oil or other organic solvent. These compositions can be made based upon light diesel fuel or kerosene. They should not be applied to crops, as they are based on oil products, which can cause plant destruction.

Baits include digestive poison containing nutrients attractive for insects. Baits are used both inside and outside the living premises.

Water soluble capsules. The capsules provide for easy, safe and economic system for pesticide application. Single dose insecticide packed within water soluble capsule can also significantly decrease transport costs. Microencapsulated insecticides together with microspheres, microemulsions and "flow" forms are the latest forms of insecticide application.

Microencapsulation helps to sharply decrease toxicity of the acting insecticide for both warm-blooded animals and humans, to eliminate bad smell of the acting substance, to prolong insecticide action on different surfaces, especially on absorbing ones, as these forms do not penetrate porous materials, but left on the surface. Insect killing active substance is contained within microcapsules i.e. polymer balls of 18 micron size on the average (Figure 45).

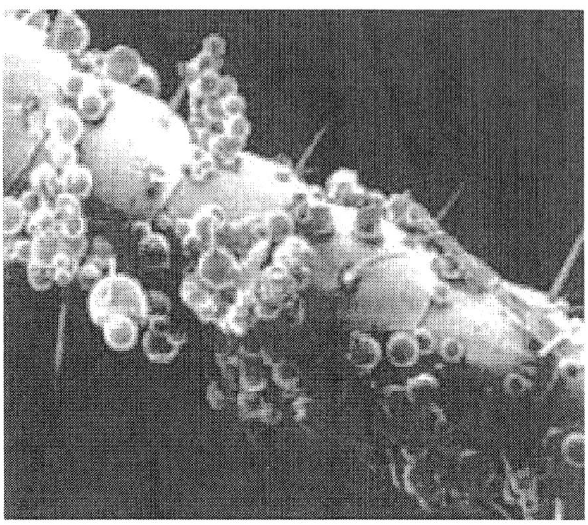

Figure 45. Microcapsules on cockroach antenna.

Wettable powder is a dry powder containing active ingredient, diluting agent (usually plastic clay or silica gel) and surfactant species. To avoid aliquation, the container with the solution to be dispersed is to be agitated.

Dry fluid suspension is a dry composition consisting of herbicide impregnated minute granules easily dispersed in water. Dry fluid powders are easy to discharge and they have smaller granules than wettable powders. To avoid aliquation, the container with the solution to be dispersed is to be agitated.

Pills are dry compacted compositions in the form of tablet impregnated with active ingredient. Pills are put in places of application or in special glue traps like houses or plates.

Runny paste is a viscous mass with a composition similar to solution containing active ingredient, water, anticoagulants and stabilizing agents contributing to maintaining active substance in suspension state. The mass should be agitated continually to avoid aliquation.

Emulsifiable concentrate (emulsion concentrate) is a fluid composition containing different emulsifiers contributing to dispersion of active ingredients in water. The weak point of the concentrates is that they can damage plants under certain weather conditions due to solvents they contain.

Table 5 presents information on application of different preparative forms of insecticides and acaricides in 1990 and 2000. Emulsion concentrates appear to be the most accepted form of insecticide application, which can be explained by easy handling of solutions with high concentration of active substance and simple preparation of process solutions for practical application.

Table 5. Development of insecticide and acaricide preparative forms (1990 and 2000)

Preparative forms	Amount in 1990	Amount in 2000
Dust	1	-
Powder	3	2
Wettable powder	34	9
Soluble powder	2	1
Watersoluble powder	-	1
Granules	16	4
Water-soluble granules	-	1
Mineral oil emulsion	1	-
Emulsion concentrate	67	62
Suspension concentrate	3	8
Water concentrate	1	1
Aqueous emulsion	-	4
Microcapsulated suspension	1	1
Runny paste	4	4
Tablet	-	3

- During American revolution the Hessian fly *Mayetiola destructor*, or Phytophaga destructor (Figure 46) "illegally" crossed the Atlantic ocean in straw-beds of Hessian soldiers and became the main crop pest

Figure 46. Hessian fly Mayetiola destructor.

Chapter 6

6. Fungicides

Fungus diseases account for significant (up to 25–30%) crop losses, as well as decrease of quality of leather-fur items at their storage. Mold fungi easily develop on any kind of leather, but it is rather vegetable-tanned leather which is more affected by fungi. Damages caused by mold result in failures of different devices and equipment with leather details. This problem is more urgent in regions with humid and warm climates. Application of oiling and filling agents in the manufacturing process additionally contributes to mold activity in leather.

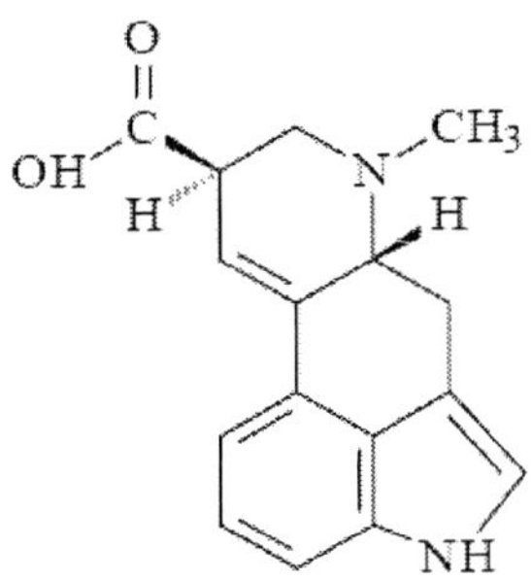

Lysergic acid

Mycotoxicosis is affection by different fungus diseases at grain growing, storage and harvesting. Black-grain and leaf smut are examples of such diseases. Fusarium fungi affect grain of all cultures mostly of true cereals. In grain mass the leaf smut is often seen as "sacs" usually of larger size and more rounded than normal wheat grains.

Content of leaf smut "sacs" (smut balls) is a black mass with bad herring odour, with their covering being fruit and seed coats of grain (Figure 47). There is no endosperm and embryos in these grains as they are fully affected by fungus. Content of leaf smut in grain is strictly monitored and if found the grain is stored and processed separately. Black-grain affects rather rye than other cereals. In grain mass black-grain is seen as sclerotia (mycelium) i.e. horns of black-violet color, 5-20 mm in length. Black-grain toxicity is stipulated by content of lysergic acid and its derivatives - ergosine, ergotamine and others with strong vasoconstrictive action. This feature of black-grain is applied in medicine for obtaining preparations for stopping the bleeding. Fungus infection takes place in the field and fungus development in storehouses stops only after decreasing the grain moisture to 14%. Grain left under snow often contains many toxins of this fungus.

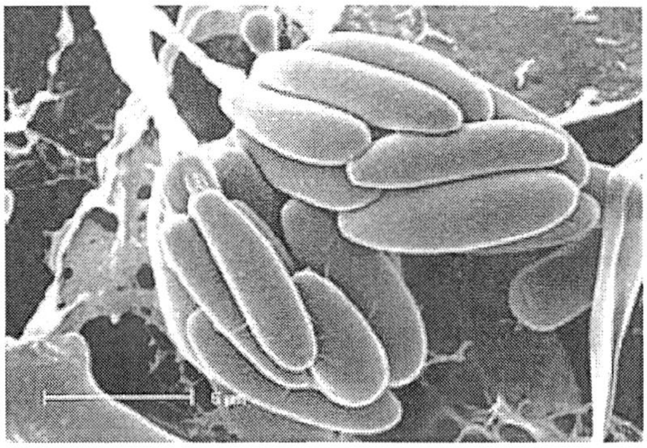

Figure 47. Black-grain sacs.

These fungi produce a number of toxins including trichothecenes and zearalenon causing severe poisoning of humans and animals. Bread baked from flour containing fusarium mycelium and eaten by humans results in poisoning looking like intoxication: there appear faintness, dizziness, emesis, sleepiness, etc. Bone marrow functions are weakening thereby resulting in sharp decrease of leucocytes in blood. Then necrotic tonsillitis appears.

Fusarium affected grain is stored separately from bread and fodder grain and use it for technical purposes. Other mold fungi also produce micotoxins which can appear on the surface of grain and products of its processing under unfavorable storage conditions. Aflatoxins affecting liver and possessing expressed carcinogenic actions are produced by Aspergillus fungi (*A. flavus* and *A. parasiticus*). Thus, the most effective way for protecting foodstuffs against fungi is exclusion of moldy grain. Smells of wormwood and smut are especially bad which do not vanish after all kinds of processing. Self-heating grain possesses malty flavor and smell; if it contained molds moldy odour is added. Lingering moldy processes result in musty and putrefactive odor.

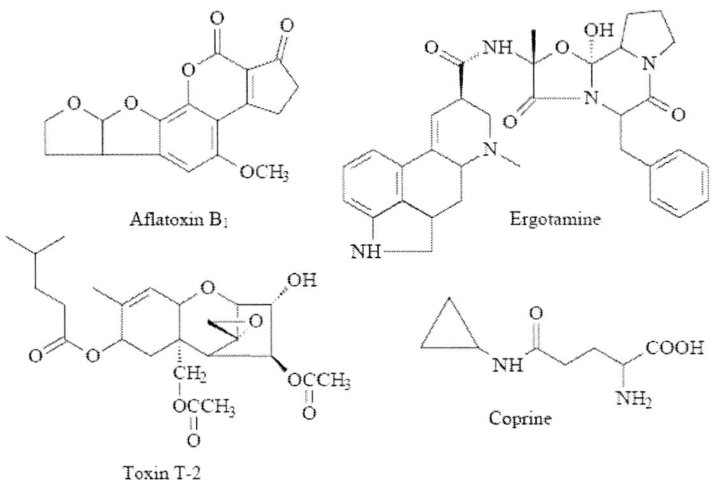

The most dangerous for human health are aflatoxin B_1, LD_{50}* of aflatoxin B_1 6.5-16.5 mg/kg. Use of foodstuffs containing 1.7 mg/kg of aflatoxin as a food for a short period of time can result in permanent liver damage, and 75 mg/kg dose – in death. Aflatoxin affection is associated with Reye's syndrome or smallpox affecting children. Symptoms are emesis, convulsions and coma. Death rate can reach 80%. Some researchers connect hepatitis B with aflatoxin supposingly changing DNA genetic organization, resulting in hepatitis virus affecting cell. Kashin-Beck disease and alimentary toxical leucopenia is the direct result of using food containing fusariotoxins. Kashin-Beck disease was originally described in the eastern part of Russia over 150 years ago. This disease is caused by fungi growing on wheat. Symptoms include brittleness of bones and bilateral osteoarthrosis* deformans. Main sources of micotoxins dangerous for humans are corn, wheat, rice, peanut and other cultures.

Poultry and other animals meat can be hazardous for human health if raised over micotoxin affected fodder. Micotoxins are very thermostable substances. E.g. ochratoxin melting point is about 220°C, aflatoxin – about 270°C, trichothecene micotoxin - 150-190°C. Besides, laboratory data and clinical examinations do not fully describe micotoxicosis, and intoxication is often misidentified as: colibacteriosis, Gumboro disease, gout, etc.

By now over 100 micotoxins have been revealed and studied; they are stable to temperatures applied at grain processing, to acids or reducing agents. The most effective way for protecting foodstuffs is mold formation prevention. Ability of etherified glucomannans extracted from yeast inner coats of specially selected strains to bind micotoxins has been discovered recently. This work lasted for almost 20 years resulted in Mycosorb making. Its composition includes no zeolites, bentonites, aluminum silicates and other sorbents. Its input ranges from 0.2 to 1 kg/t depending on product toxicity level. Besides, Mycosorb adsorbs not only aflatoxins but a number of other dangerous toxins, including T-2, DON, ochratoxin, etc.

Fungicides are substances toxic for fungi and suppressing development of their spores or mycelium. Fungicides applied for treating plants are divided into protective, systemic and medicative (exterminative). Protective fungicides are used for precautions, medicative fungicides are used for treating after pathogen* penetration, thereby suppressing symptoms of a disease in plants.

Systemic fungicides are substances capable to move along vascular system of plants and protect new shoot appeared after treating, while contact ones protect those parts of plants that are covered. In many cases systemic fungicides possess protective or healing properties, while contact ones – protective only. Systemic fungicides are quickly absorbed by the plants, so their effectiveness is less dependent on precipitations than contact ones.

Great majority of pathogens are agents of plant diseases. Accordingly, to solve all complex of problems for plant protection against diseases, in particular to minimize appearance of agent resistance, a large spectrum of fungicides and their combinations is required to be applied both continuously and alternatively. A new generation of fungicides – strobilurins – appeared recently. This class of substances has been elaborated and issued in

---

* LD_{50} – minimum dose of a substance causing death of 50% species, which are being tested for acute toxicity of preparation, from Latin *letalis* – lethal.

* Osteoarthrosis – chronic desease of the joints characterized by destruction of articular cartilage and appearance of structural changes in the bone surrounding joint.

* Pathogen (Greek pathos – anguish + Greek genos – origin) are pathogenic or parasites causing infectious diseases.

the market to increase efficiency of plant protection against main diseases such as false mildew, oidium, buck eye rot, rhynchosporiose, and eyespot.

Externally simple edible fungus *Strobilurus tenacellus*, looking like toadstool (Figure 48) grows on fallen pinecones and possess high fungitoxic activity, as it releases substances which suppress growth and development of other fungi. Other fungi are destroyed as competitors by feeding.

Figure 48. Strobilurus tenacellus.

The fungus disperses these substances around and suppresses microscopical spores bit its action does not appear to be strong and prolonged due to high light sensitivity. BASF company chemists have made synthetic strobilurin ecologically safe-health (LD_{50} for strobilurin A is exactly 500-825 mg/kg [60, 61]]) and environmentally safe.

New fungicide solution is evenly dispersed over the surface of leaves and fruit as if covering them into protective shell against rain, or either high or low temperatures. The preparation is readily imbibed by water and is economic one. Only 2 grams is required per 10 liters of water. Based upon A strobilurin it was elaborated cresoxymethyl, which is a reactant of "Strobi" – a fungicide of new generation safely protecting plants and environmentally safe (water, animals, soil).

Unlike other preparations "Strobi" penetrates through a leaf and evenly spreads over the surface and inside it, even if only one side of a leaf is treated, with the plant absorbing just smaller part of reactant. The unique feature of the preparation is its high rain resistance. Weak and mean rainfall (up to 40 mm) does not wash away the reactant, because it is absorbed into wax layer of leaves and fruit. High rain resistance makes it possible to spray wet leaf areas also. "Strobi" does not kill beneficial insects (bees, predatory mites, etc.) and is extremely temperature-stable and weatherproof [62].

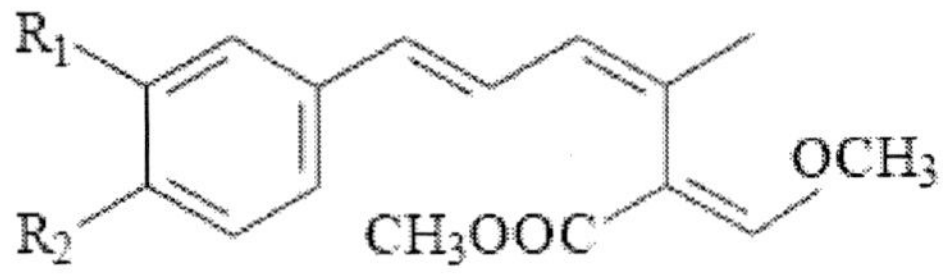

$R_1 = R_2 = H$ (Strobilurin A or mucidin);
Strobulurins:
$R_1 = OCH_3$, $R_2 = H$ or Cl;
$R_1 = H$, $R_2 = OMe$; $R_1 = OH$, $R_2 = H$;
$R_1 = (CH_3)_2C = CHCH_2O$, $R_2 = H$;
$R_1 = OH$, $R_2 = (CH_3)_2C = CHCH_2O$

Myxothiasol extracted from bacteria possesses structural fragments as β-methoxyacrylic acid, included into strobilurin A.

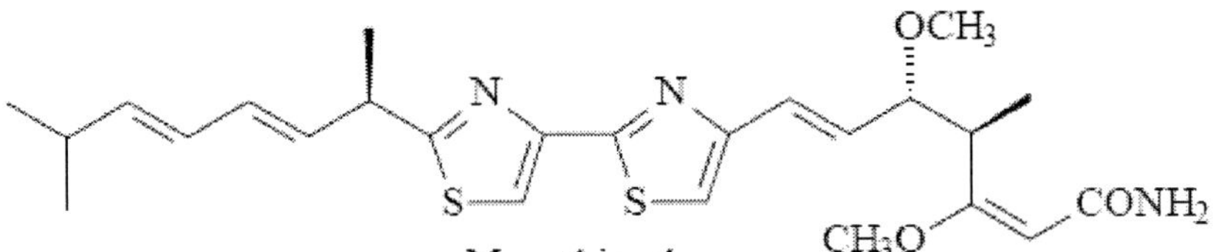

Myxothiasol

Mechanism of action of these compounds is similar as well i.e. cell respiration inhibition in the place of complex of electron carrying cytochromes bc_1. Chain of respiratory enzymes carrying electrons plays an important role in cell energy-supply of all aerobic organisms, including bacteria and fungi. Synthetic analogues of natural strobilurins A ICI-A5504 and BASF-490F have been patented as fungicides of wide spectrum of action, nematocides, insecticides and acaricides which application is promising both in agriculture and in industry.

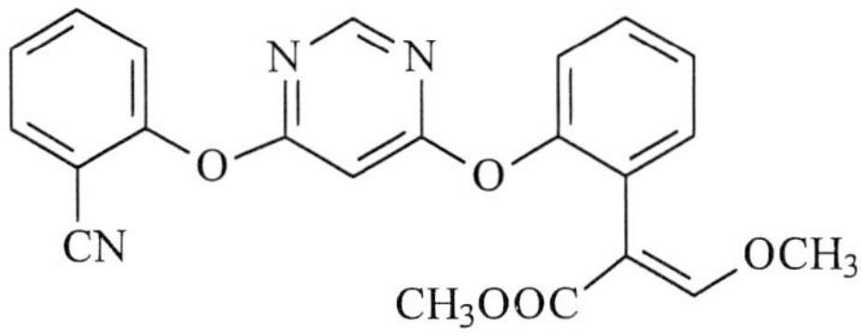

ICI-A5504

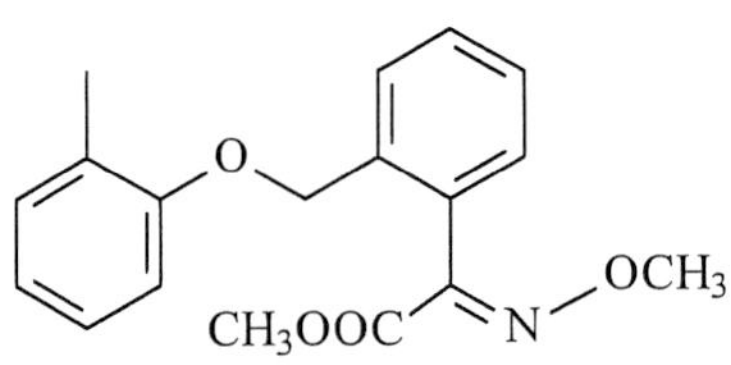

BASF-490F

Synthesis of strobilurin analogues is presented in literature by several versions. In one of them substituted acetic esters *1.* were used as original compounds. These esters either interact with formats or result in aldehydoester of type *2* through intermediate adduct with dimethylformamide dimethyl acetal. Alkylation of aldehyde enolates *2* gives β-alkoxyacrylates [62]:

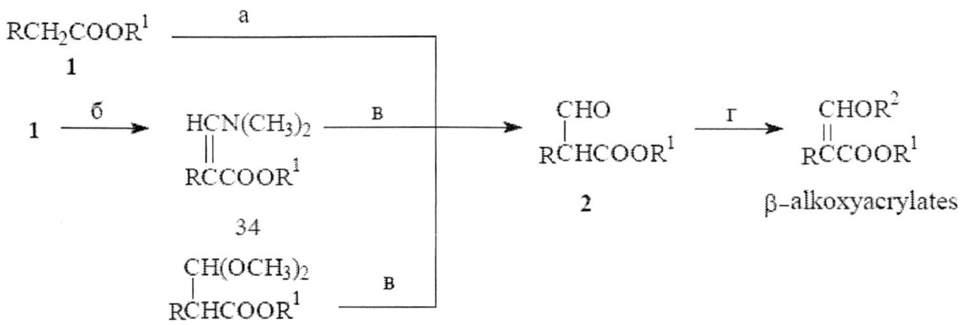

Reagents: a.HCOOR, NaH; б.$(CH_3)_2NCH(OCH_3)_2$, PyTsOH; в.H_3O^+;
г. H_3O^+, R^2X (X= Hal, SO_3OR^2); R^1,R^2 = Alk; R = Ar, Het, ArX-, HetX- etc.

β-methoxyacrylates are often synthesized by Wittig reaction proceeding from glyoxalic acid esters 3 [63, 64]:

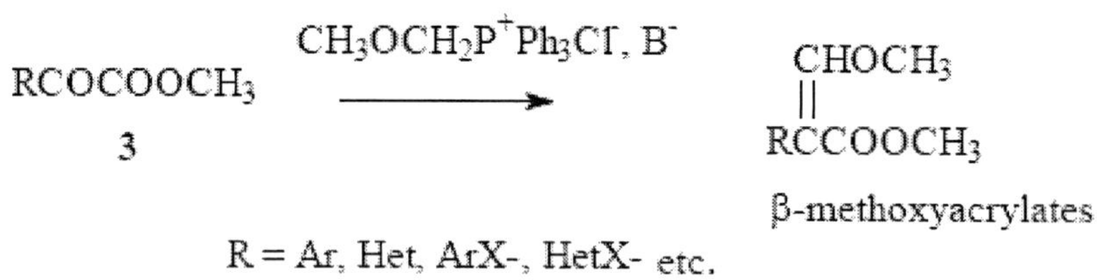

Nowadays fungicides are mainly applied for protecting fruit, green crops, cereals and rice. These preparations are more important for cereals in high farming, when the crop capacity increases together with the absolute value of losses, furthermore, the more fertilizers are introduced, the less is the plant resistance against pathogenic organisms. Besides, disease resistance of plants is also influenced by weather conditions, in particular, systemic fungicides and protectants are applied for disease prevention under high humidity and continuous rainfall.

There is another highly effective fungicide of wide action spectrum against diseases of eared cereals – propiconasole. It is effectively protects against such diseases as, oidium, brown, stem and stripe rust, barley net blotch, winter wheat leaf blotch, scald, cladosporiosis, etc.

Propiconasole belongs to triazole fungicides of systemic action. Propiconasole fungistasis is suppression of ergosterol*, due to which cytoplasmic membrane structure of fungic cell is lost, and the cell cannot grow and develop. Squalenes are accumulated within the cell, which like lipidic sponge, extract lipidic components from cell membrane thereby impairing its properties. Lipid granules being accumulated within the cell are gradually

* Ergosterol – steroid precursor of vitamin D.

expanding and finally break cytoplasmic membranes resulting in fungic cell death. The preparation is highly selective to all treated cultivated plants. It belongs to low-toxic pesticides, LD_{50} for rats is 1517 mg/kg. The preparation does not irritate skin and eye mucosa. It is actually safe for bees, birds and fish, and beneficial entomofauna. The preparation is easily decomposed in soil: half-life period is 41-70 days.

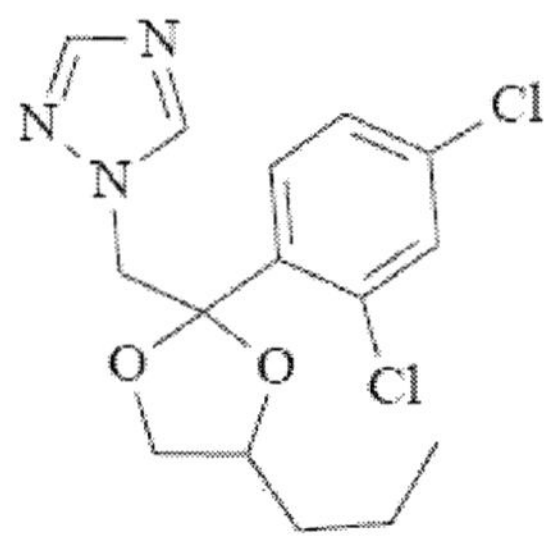

Propiconasole (tilt)

Imidazole derivatives are widely used in agriculture as fungicides. The most general technique for getting imidazole and its derivatives **4** is based upon condensation of dicarbonyl compounds with ammonia and aldehydes:

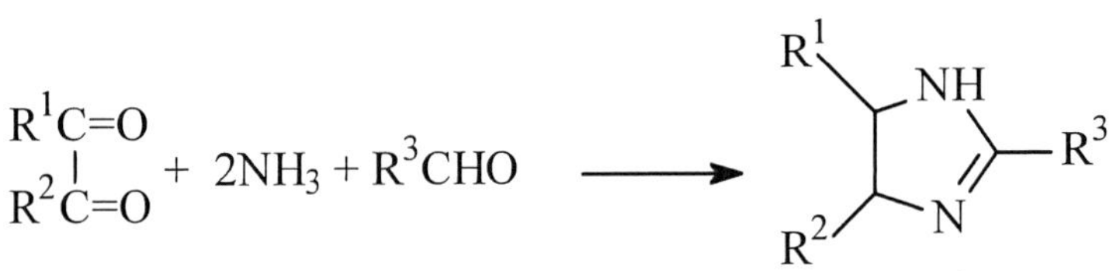

4

Carbendasim (N-[1-(N-butylcarbamoyl)benzilimidazolyl-2]-O- methylcarbamate) is effectively suppressing fusaruim mold, Cercosporella spot, and is used for seed treatment and applied for control against fungi *Cladosporium*, *Aspergillus* and *Penicillium*, affecting leather semi-products.

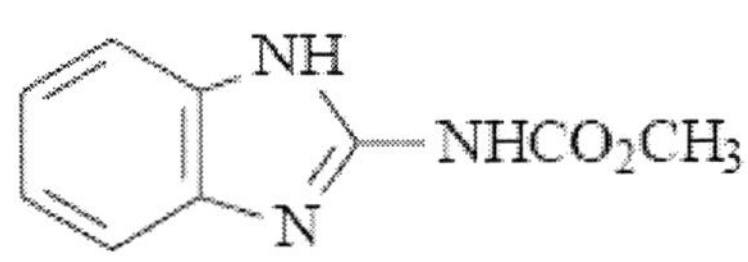

Carbendasim

Along with carbendasim tanning industry uses 2-mercaptobenzothiazole (captax, MBTS), 2-*н*-octylisothiazolin-3-one (Kathon LP) and 2-(tiocyanomethyltio)benzothiazole (Busan 30A, TCTMB). Table 6 gives comparative data on minimum doses suppressing growth and development of wide spectrum of mold species, affecting leather semi-products. From the Table we can see that Kathon LP is essentially exceeding Busan 30A in its activity on all main species of molds affecting leather semi-products.

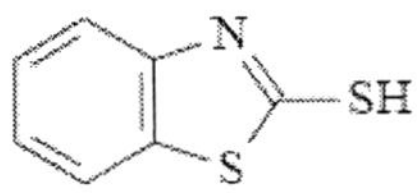

Captax

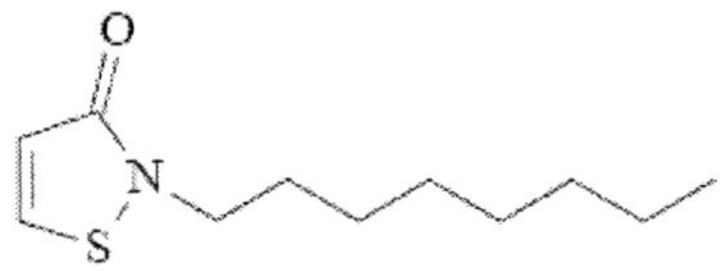

Kathon LP

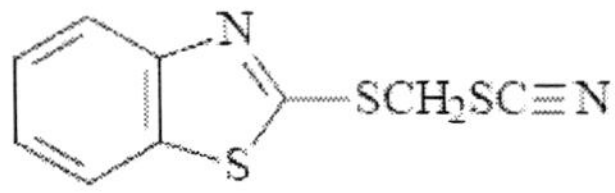

TCTMB (Busan 30A)

Table 6. Activity of Kathon LP and Busan 30A on molds

Fungi	Kathon LP, ppm*	Busan 30A, ppm
Alternaria alternata	4	125
Aspergillus niger	8	125
Aureobasidium pullulans	0,5	375
Chaetomium globosum	4-10	50
Gladosporium resinae	0,5	NA
Penicillium funiculosum	1	NA
Penicillium glaucum	2,5	100

2-(thyazollil-4)benzylimidazole (thiabendazol or tecto) are used for treating granaries and seeds.

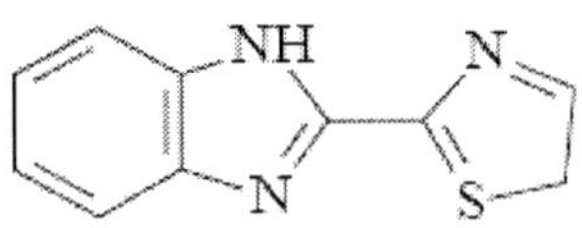

Thiabendazol

Dithiocarbamic acid derivatives are applied as fungicides of wide spectrum of action for integrated plant protection system. These preparations are effective against diseases of potato, apple-, pear- trees, green crops, etc.

Preparations based upon ethylene-bis-dithiocarbamates are of special importance among them. They include Polyram DF, water-soluble granules (w.g.) issued by BASF AG (Germany); Pencozeb, wettable powder (w.p.) of CerexAgri company (France); Ridomil Gold MZ WP, w.p. and Ridomil Gold MZ WG, w.g., issued by Syngenta (Switzerland). Pencozeb is a powder of sulphur color, odorless and it contains 800 g/kg of mancozeb as

*ppm - the dose is expressed as amount of tested substance present in unit of air volume: mg/m^3 or parts per million (ppm - parts per million).

reactant. It is applied for control against late blight disease and macrosporios of potato, and against mildew* in vineyards. Mancozeb is [[1,2-ethanediylbis[carbamodithioato]](2-)]manganese mixture with [[1,2-ethanediylbis[carbamodithioato]](2-)]zinc:

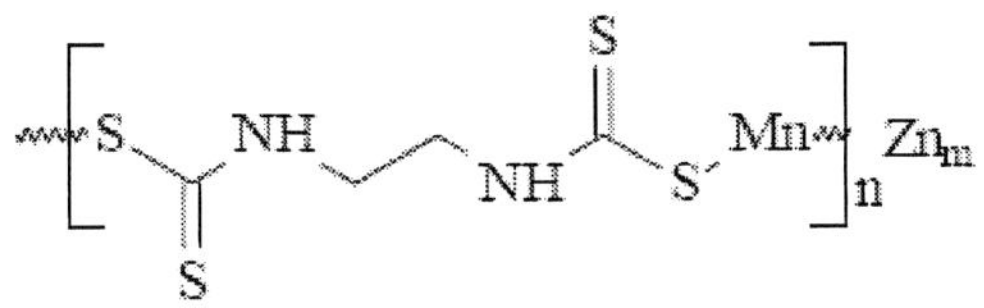

Mancozeb

Maneb is synthesized by interaction between ethylene-N,N[1]- bis(dithiocarbamate) of ammonium (5) and manganic sulphate in water solution:

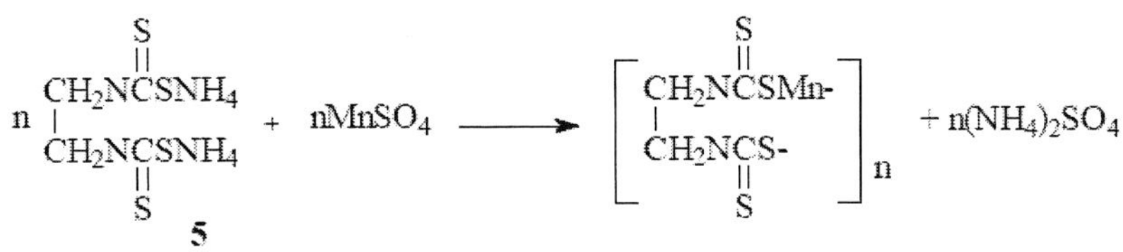

Maneb

Mercury compounds are also used for seed treatment as a fungicide. In Sweden after starting wheat seed treatment by methylmercuric dicyandiamide mercury concentration in seed grain made 15-20 mg/kg. In Iraq bread made of mercury treated seed grains resulted in mass poisoning of population [65].

Copper and chromium compounds (III) fungicidally affect *Fusarium* molds parasitizing on plants and skin collagen.

Table 7. Indicators of fungicide acute toxicity based on ethylene-bis-dithiocarbamate

Indicators	Methiram	Mancozeb	Polyram DF, w.g.	Pencozeb, w.p.	Ridomil Gold MZ WP
LD_{50} (mg/kg) *per os*, rats	6500	>5000	>5000	>5000	>2000
LD_{50} (mg/kg) cutaneously, rats	>5000	>2000	>2000	>2000	>4000
*LC_{50} (mg/m^3) rats	>5700	>4760	>2710	>4100	—

Rawstock and fur stock are received by processing plants in microbiologically seeded state. Moisture and dirt in workshops are ideal conditions for fungi growth and development.

* Mildew (false mildew) – disease of grapevine caused by parasitic fungi.

* LC_{50} – lethal concentration.

Phenolic compounds such as 4-chloro-*m*-cresol, *ortho*-xenol, 2,4,6-trichlorophenol and pentachlorophenol were used 10 years ago in tanning industry.

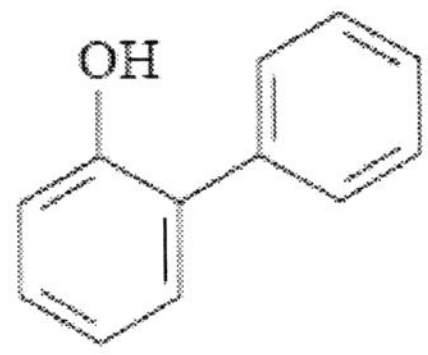

ortho-xenol

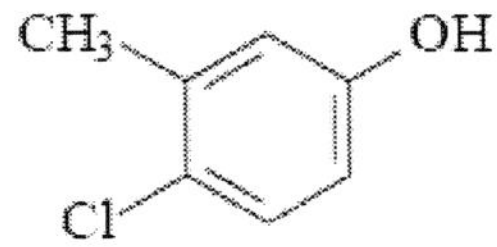

4-chloro-meta-cresol

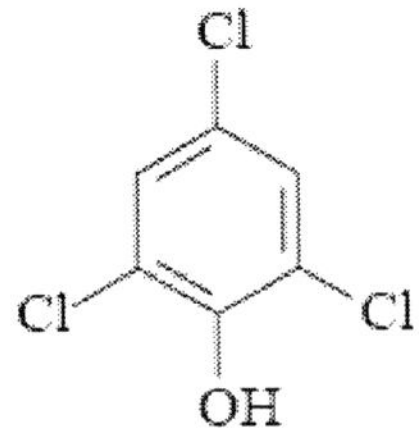

2,4,6-trichlorophenol

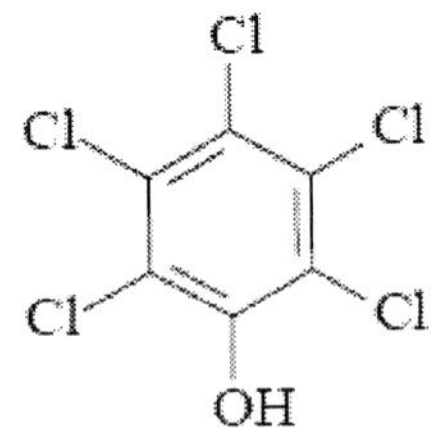

Pentachlorophenol

However at present application of pentachlorophenol and other chlorinated phenols is prohibited actually in all European countries and the USA. It is mainly stipulated for chlorinated phenols are considered to be reasons for carcinogenic and high toxic dioxins formation.

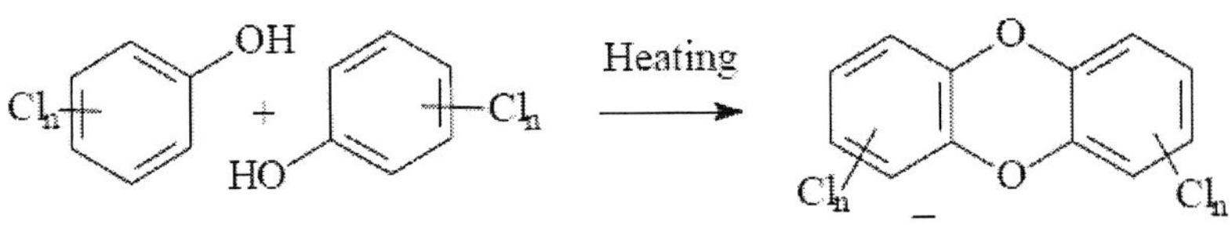

Dioxins

Chapter 7

7. Organochlorine Insecticides

Organochloride pesticides are a great class of chemical substances including halogens of alicyclic and aromatic compounds. They are usually solid crystalline substances with high thermostability, low pressure of saturated vapor ($10–4 \cdot 10^{-7}$ mmhg), resistance to concentrated acids, and bad water solubility. The most popular organochloride insecticide is 4.4′-dichloro-diphenyl-trichloroethane. Although this substance was synthesized still in 1874 its insecticide properties were found in 1939 only by a Swiss chemist Paul Muller, at the same year this preparation was patented by Geigy corporation later known world wide as DDT insecticide. In 1940 it was tested and in 1942 it came into the market. And in 1948 Swiss chemist Paul Hermann Muller (1899-1965) was awarded the Nobel prize in physiology and medicine for synthesis and research of DDT pesticide properties.

In Russia the plant protection research was made by Academician A.N.Nesmeyanov in the laboratory for chemical protection of plants established in early 20s of the XX century. Later in 1929 this laboratory was reorganized into All-Union Research Institute for Chemical Protection of Plants (AURICPP) (since 1931 - Insecto-fungicide Institute), and since 1933 – Research Institute for Fertilizers and Insecto-fungicides (RIFI). Immediately after the Patriotic War the latter institute has started systematic research in organochloride pesticides, including DDT and its analogues like hexachlorocyclohexane, diene synthesis preparations based on hexachlorocyclopentadiene. Scientists from this institute elaborated technologies for making chlorinated turpentine being the first domestic organochloride insecticide produced commercially. Later it was made DDT preparative form like antiparasitic pencil, produced by experimental plant at AURICPP in Schelkovo.

Other more effective preparations were created among which the most popular were chlorine organic compounds: hexachlorocyclohexane or its gamma isomer lindane, chlordane, dieldrin, aldrin and toxaphene (polychlorocamphene):

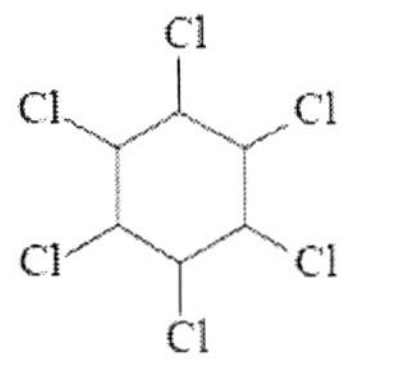
Hexachlorocyclohexane

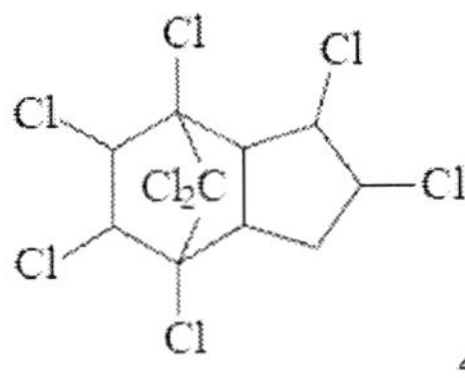
Chlordane

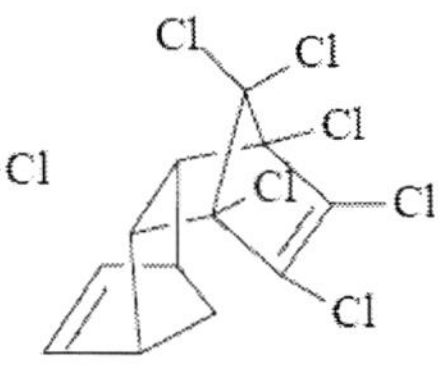
Aldrin

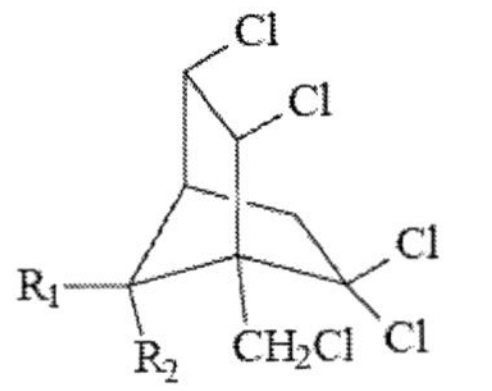
R_1=CH_2Cl, $CHCl_2$; R_2=$CHCl_2$, CH_2Cl

Toxaphene

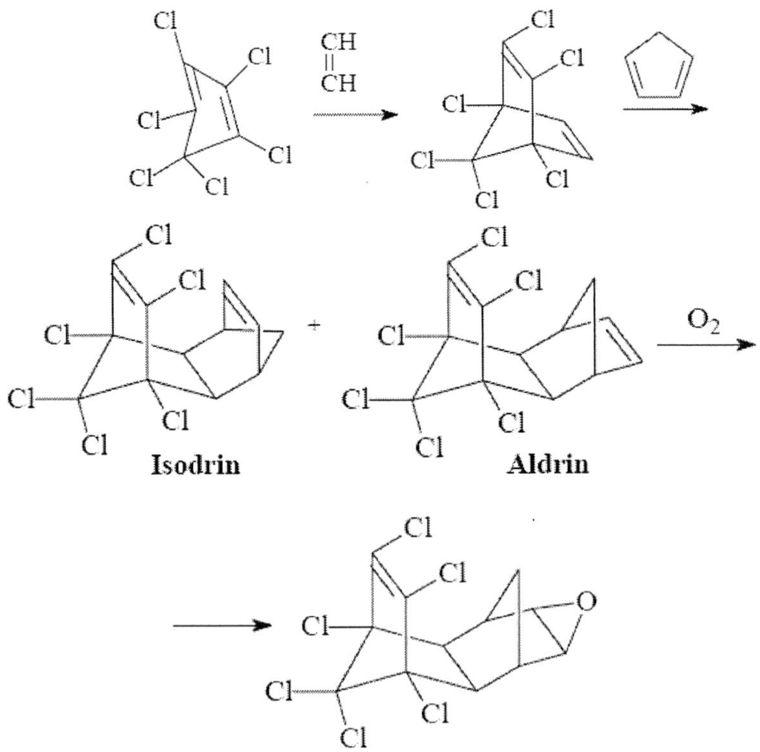

Main disadvantages of organochloride insecticides are their toxicity and absence of selective ability. All insects are anyway sensitive to these preparations, thereby resulting in death of all species in the treated area including beneficial insects as well. In France, it was found that 48% of Diptera species, 21% of Hymenoptera, 14% of Coleoptera, 15% of Hemiptera and 2% of butterflies were killed in great area treated with hexachlorocyclohexane against chafers. Great majority of insecticides is very dangerous poison for beneficial animals in farming. Thus, while killing wigglers and buffalo gnats cold-blooded animals are also affected. Some insecticides like DDT are disastrous for fish, with their sudden death being caused by damage of the central and peripheral nervous system, respiratory system and disorders in separate spheres of metabolism; at low doses harmful effect of these insecticides is expressed in major physiological disorders [66]:

A majority of polychlorine derivatives are stable compounds, very slowly decomposing in nature and accumulating within living organisms. Large scale of DDT application is largely explained by its easy synthesizing. Basic method for getting DDT is chlorbenzene condensation with trichloracetaldehyde over strong sulphuric acid or oleum. Chloral is synthesized by chlorination of ethyl alcohol over water:

$$C_2H_5OH + 3Cl_2 = Cl_3CCHO + 3HCl + H_2O$$

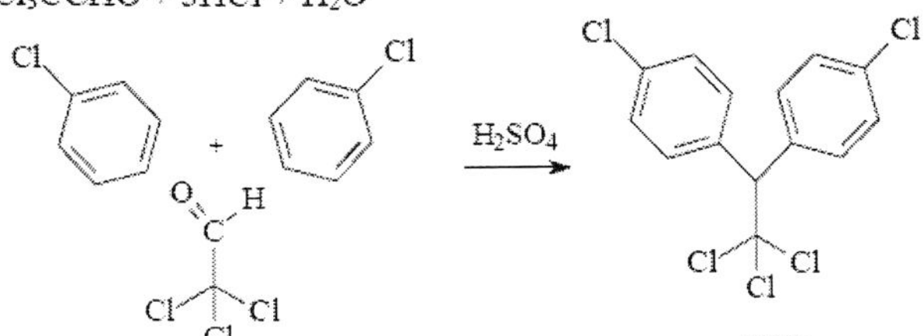

DDT

Treatment of wetlands and mountain forests rich in water flows with organochloride insecticides is strongly affecting the balance of water dwellers, especially fish. DDT spraying from the air resulted in death of hundreds of thousands species of trouts and salmons in western parts of the USA and Canada, while insects – forest worms appeared again after each treatment. Nowadays DDT application is prohibited in Russia and in some other countries, however, it is still apllied in some countries of Asia and Africa.

Serious illness like filariasis is onchocercosis. This parasitic disease is caused by filarial (*Onchocerca volvulus*) penetrating tissue transmitted by black flies *(Simulium damnosum,* small Diptera which larvae can be found in hydrophytes in rapid current rivers. Under onchocercosis the cyst is formed in the eye, the eye functions are disturbed and the patient is becoming blind. There are no preventive measures against onchocercosis, and modern medicine achievements are still poor in this field. Principal method for fighting this disease propagation is extermination of carriers of disease – black flies. Starting since 1955 vast areas around African Haute-Volta river have been systematically and repeatedly in certain intervals treated by lindane. However, these disinfestation measures failed to kill carriers of disease, but resulted in elimination of fish stock and undermined food economy in these regions.

Aromatic organochloride insecticides noted for their high resistance to environmental decomposition affect birds, mammals and also man as the last link in trophic food chain. Lethal doses of some of these preparations are relatively small (LD_{50} of dieldrin is 20-30 mg/kg, endrin kills the rat at 7.5 mg/kg dose). Treated insects after being eaten by insect-eating birds can poison the latter ones with dosage toxic both for them and their nestlings.

The most harmful was seed treatment with insecticides (aldrin, dieldrin and heptachlor were mainly applied against some insects, and mercuric compounds were used against fungus diseases). It should be taken into account that insecticides are easily concentrated in sex glands of birds (with their concentration being here 30-211 mg/kg), which results in partial or full sterilization of begetters.

To control stored grain pests aliphatic series halogens are applied: methyl bromide (CH_3Br) – colorless liquid with boiling-point of 3.6°C, it is high-toxic substance for both humans and animals. Possessing high reactivity methyl bromide is able to negatively affect grain properties, methylating some of its components and changing its properties and thereby reducing its germinability.

Chapter 8

8. Organophosphorous Insecticides and Fumigants

Gerhard Schroder formerly employed by Bayer Chemical company in 30s of the XX century was regarded as a pioneer in synthesizing organophosphorous pesticides. Along with elaboration of organophosphorous pesticides on the eve of the WWII this company has made nerve agents: sarin, soman and tabun.

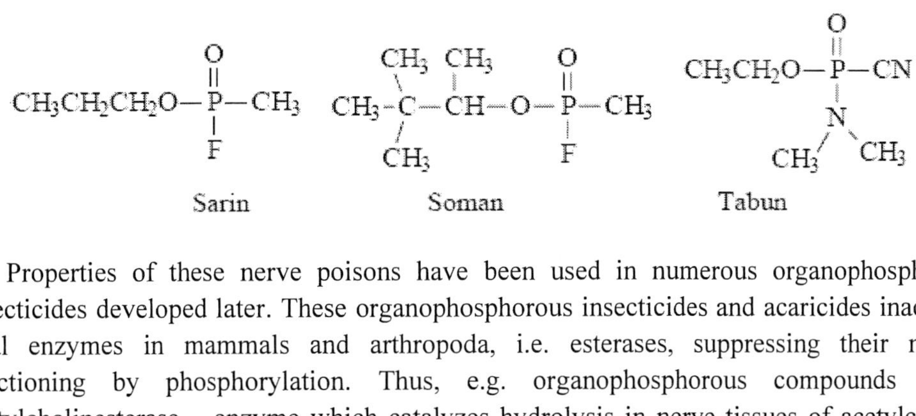

Sarin Soman Tabun

Properties of these nerve poisons have been used in numerous organophosphorous insecticides developed later. These organophosphorous insecticides and acaricides inactivate vital enzymes in mammals and arthropoda, i.e. esterases, suppressing their normal functioning by phosphorylation. Thus, e.g. organophosphorous compounds affect acetylcholinesterase – enzyme which catalyzes hydrolysis in nerve tissues of acetylcholine being transmitter of nervous impulses [67].

Dichlorvos belongs to this class of organophosphorous insecticides. This is one of phosphoric acid derivatives characterized by high toxicity both for humans and animals. For a long time it was used in sprays for synanthropic insects control. Dichlorvos can be synthesized by reaction of chlorophos dehydrochlorination *6* under alkali action:

6

Dichlorvos

Crufomate is applied to control oxflies seriously damaging domestic animal hides thereby decreasing rawstock qualities, animals weakening and milk-yield fall. It is used as

0.25-0.7% solution, it is not accumulated in animal organisms and quickly environmentally decomposed. It is synthesized by interaction between 2-chlor-4-*tret*-butylphenolate of sodium (7) and *O*-methyl-*N*-methylamidochlorphosphate (*8*).

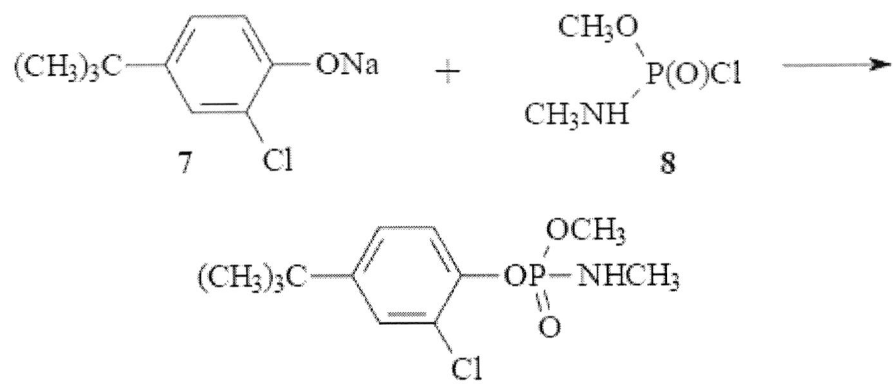

Crufomat, LD_{50} = 660-1000 microgram/kg perorally for rats

- Great damage to horse-breeding is caused by botflies. The most typical disease caused by bots parasitized within the horse is gastrophilosis. Botfly female lays its eggs to root part of head hair. Matured larvae drill into skin and complete development in stomach or intestine. Larva activity can be strongly harmful for the horse organism, even can result in enterobrosia. Under strong invasion• the horse is becoming weak, loses appetite and work capacity. In severe cases up to 1800 larvae of 10-12 mm in length have been found in horse intestine. It is very difficult to diagnose that case, however, under rectal examination larvae attaching rectum can be found. Insecticide treatment basically applying carbon tetrachloride and chlorophos is quite effective. Disease prevention is in treating horses with chlorophos emulsion before depasturing. Botflies cause other diseases as well: rhynoestrosis (larvae parasitizing in nasal cavities) and hyposermosis (larvae development under skin). Prevention methods are the same – animal treatment before depasturing.

Insecticides which are low toxic for warm-blooded animals are found among derivatives of thiophosphoric acid, including phoxim (baition, volaton), actellic (pirimiphos-methyl) and diazinon (basudin) applied as emulsifiable concentrates, wettable powders, granules or for fumigation of storehouses against corn ground beetle, weevils and other pests.

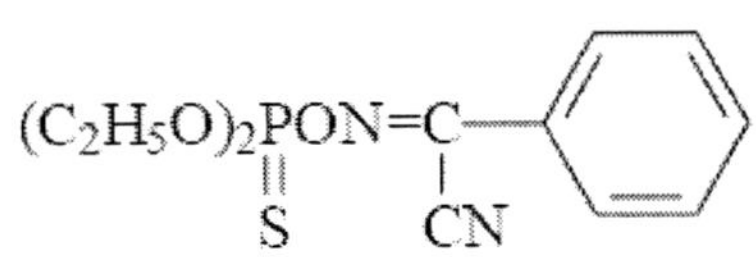

Phoxim, LD_{50} = 200 microgram/kg perorally for rats

• Invasion (from Latin. *invasio* attack, intrusion) - the act of invading the organism of humans, animals or plants with animal parasites (protozoa, worms, arthropoda) with further development of complex pathophysiological process of their interaction.

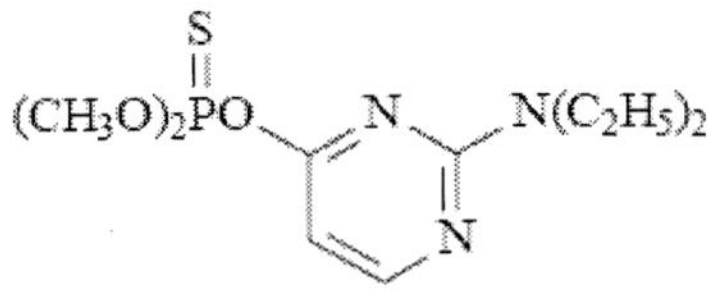

Actellic, LD_{50} = 2050 microgram/kg perorally for rats
Insecticide and acaricide for granaries protection

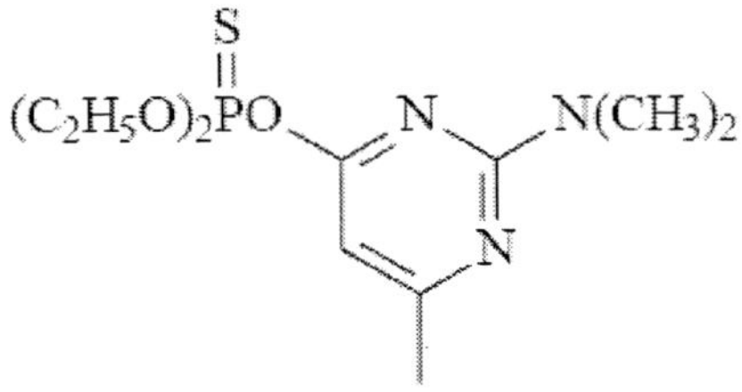

Diazinon, LD_{50} = 120-220 microgram/kg perorally for rats
Insecticide and acaricide for granaries protection

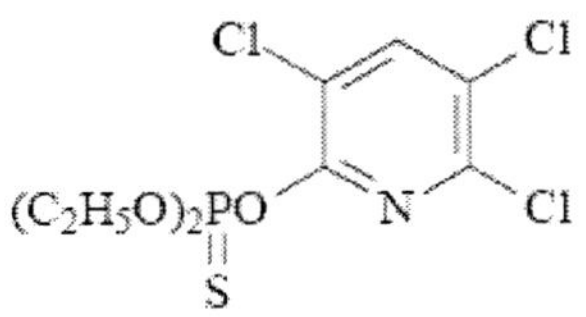

Chlorpyrifos, LD_{50} = 25000 microgram/kg perorally for rats

Reactant of EMPIRE®20 preparation making efficient insecticidal activity on wide range of domestic insects is chlorpyrifos (see Table 8).

Table 8. Insects affected by chlorpyrifos

Insects	Species	Latin names
Cockroaches	American, brown-banded, German, common, Asiatic	*Periplaneta americana, Supella longipalpa, Blattela germanica, Blatta orientalis, Blatella asahinai*
Ants		Formicidae spp.
Dermestids		Dermestidae spp.
Mites		Ixodidae spp., Argasidae spp.
Fleas	cat, dog, Oriental, rat, human	*Ctenocephalides fells, Ctenocephalides canis, Xenopsylla cheopsis, Pulex irritans*
Crickets		Gryllidae spp.
Earwigs		*Dermaptera spp.*

Table 8. (Continued)

Insects	Species	Latin names
Bristletails, silverfish		*Thysanura spp.*
Flour beetles	confused, red	*Tribolium confusum, Tribolium castaneum*
Grain beetles		Oryzaephilus surinamensis
Granary and maize weevils		*Sitophilus spp.*
Indian meal moth		*Plodia interpunctella*
Flour pyralid moth		*Pyralidae spp.*

Carbophos being thiophosphoric acid derivative is widely applied as insecticide. Dithiophosphates unlike thiophosphates are less toxic and more resistant to decomposition in field conditions. Carbophos within typhoid fly organism influenced by specific enzymes is transformed into more toxic diester 9, while in rat organism malathion acid S-(1.2-dicarboxyethyl)-O.O- dimethyldithiophosphate (10) low-toxic for animals is formed.

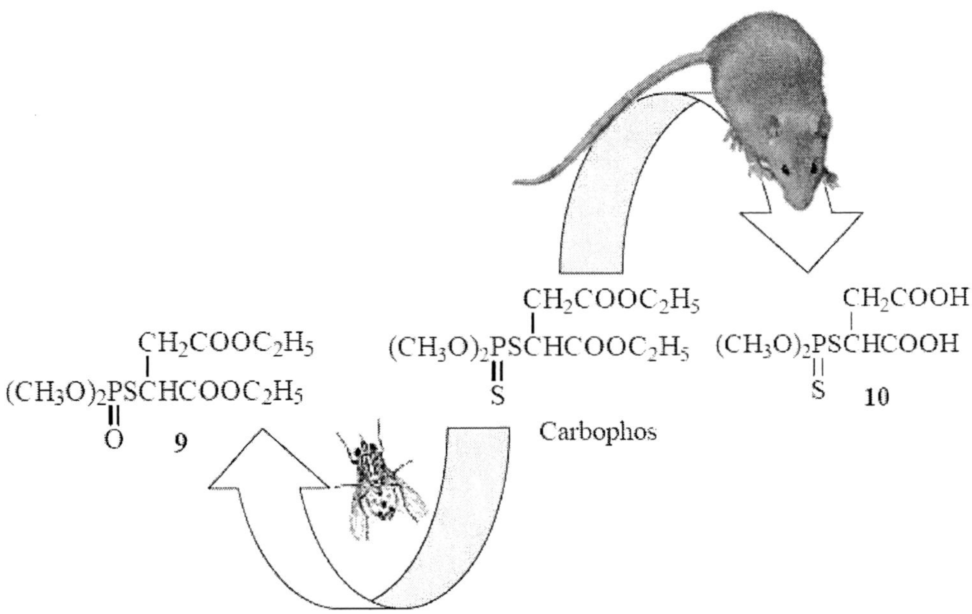

Carbophos is formed by interaction between dimethyldithiophosporic acid (11) and maleic acid ester (12):

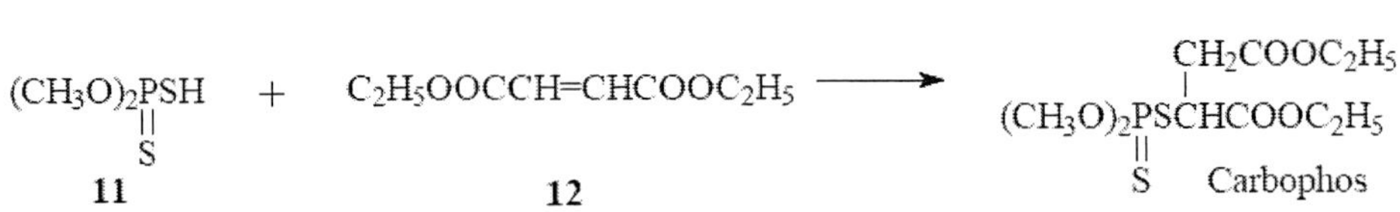

The reaction takes place over catalysts with or without solvents. Technical carbophos obtained from refined dimethyldithiophosporic acid is pure enough and contains up to 4% of impurities. When contacting iron for a long time carbophos is decomposed and loses its insecticide properties. Long heating results in carbophos isomerization forming more toxic isomer 13:

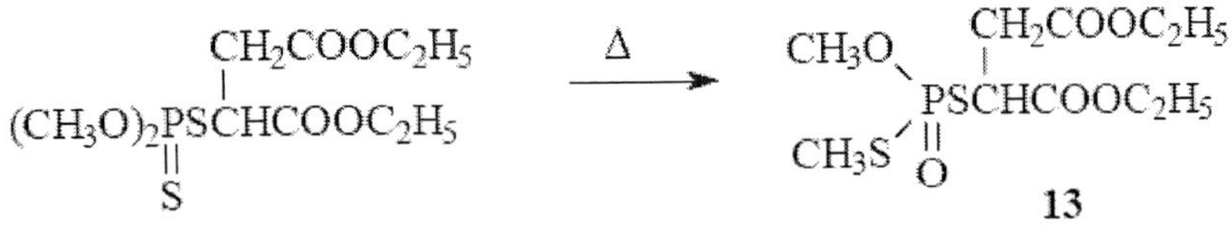

13

Chloride of (3.4-dichlorobenzyl) triphenyl phosphonium (**14**) is part of preparation for treating wool against moth.

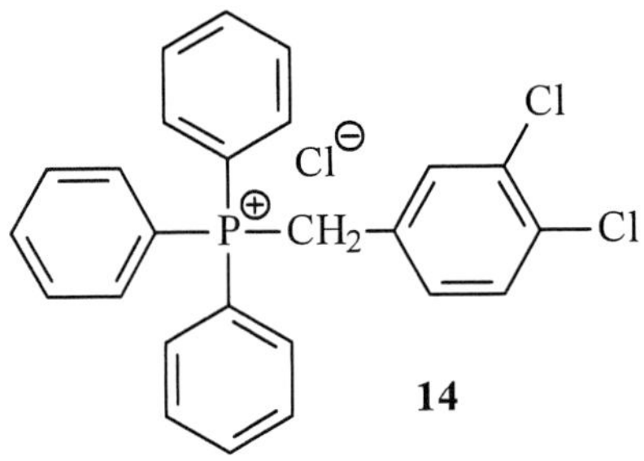

14

Chapter 9

9. Carbamates

Some of carbamates (NH_2COOR, where R is hydrocarbon radical) occurring in nature are applied as insecticides. Carbamates are esters of carbamic acid NH_2COOH not occurring in a free state and usually they are crystalline substances. Originally carbamate insecticides were extracted from calabar-beans (Figure 49) growing in West Africa in body of rivers Niger and Calabar.

Figure 49. Calabar-beans *Physostigma venenosum.*

According to ancient customs a person sentenced to death was to eat calabar-beans by order of a tribe chief. The tested considered himself not-guilty quickly ate them. Toxic substances after reaching stomach resulted in emesis (cholinergic response) and this saved his life. Guilty person paralyzed by fear was eating beans slowly delaying the punishment. Thereby he was eating fatal dose not resulting in saving emesis when eaten by small parts.

One of main toxic substances in calabar-beans *Physostigma venenosum* is physostigmine which was extracted in 1865, with its structure being defined in 1925. It was synthesized in 1935 and trivially named as eserin.

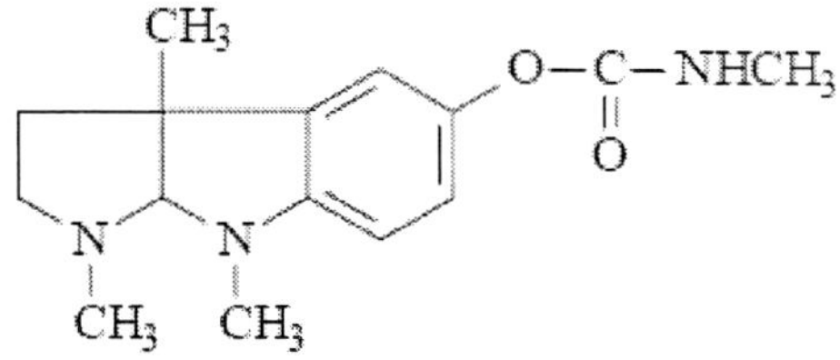

Physostigmine (eserin)

Later many different analogs of natural carbamats were synthesized.

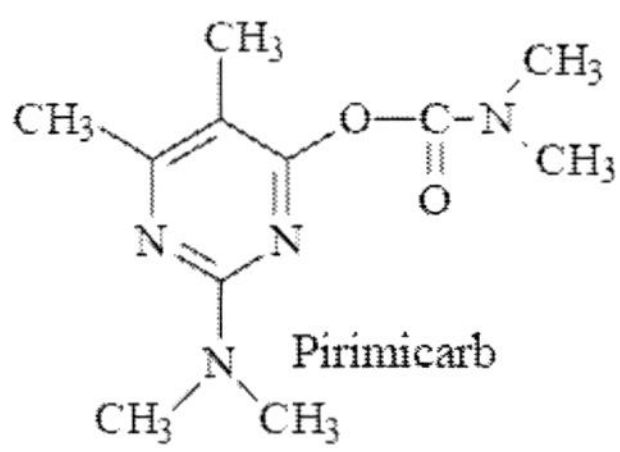

Pirimicarb

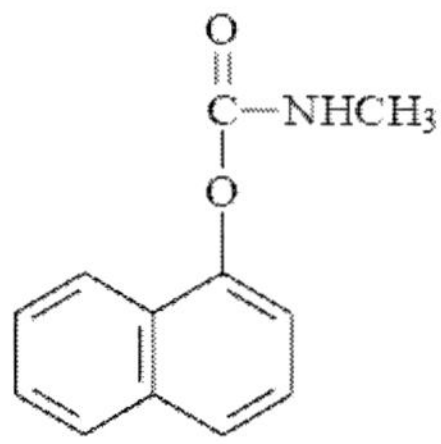

Carbaryl

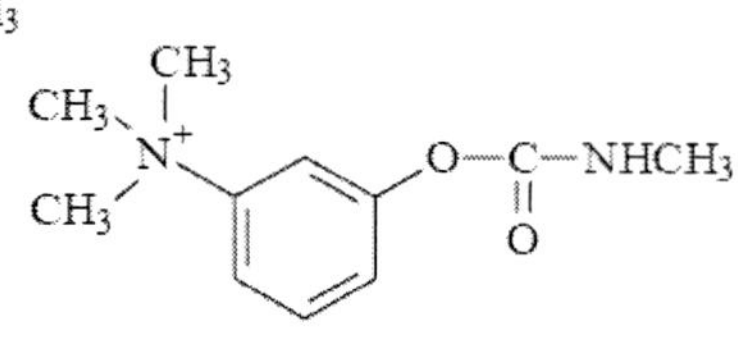

Neostigmine

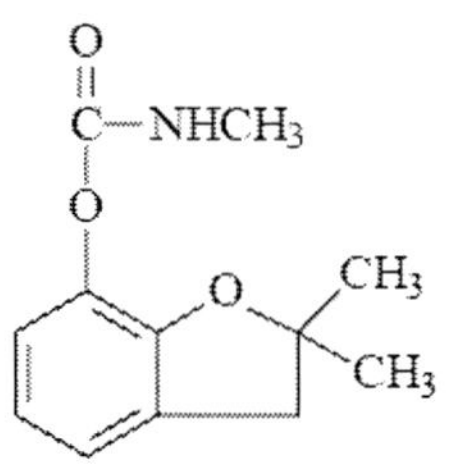

Carbofuran (furadan)

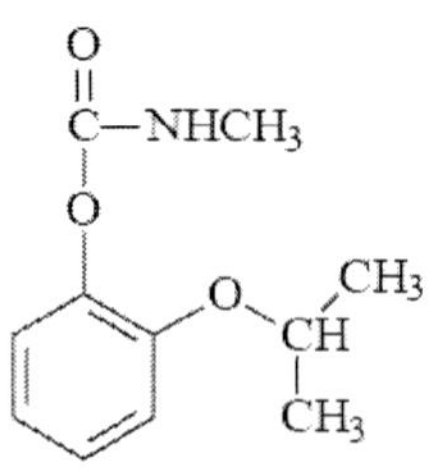

Propoxur (baygon)

Among synthetic carbamates the most effective is and was carbaryl obtained from α-naphthol and methyl isocya almost with quantitative output.

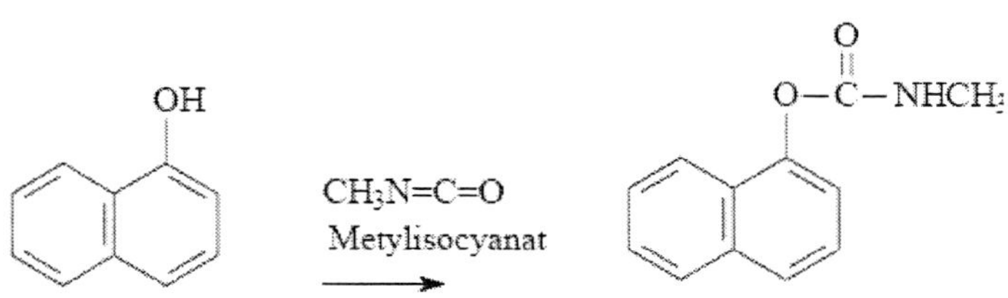

α-naphthol Carbaryl

Carbamate toxicity for mammals and insects is stipulated by formation of complex of enzyme (acetylcholinesterase AChE) with carbamate, by blocking activity of cholinesterase enzyme and thereby stopping nervous impulse transfer.

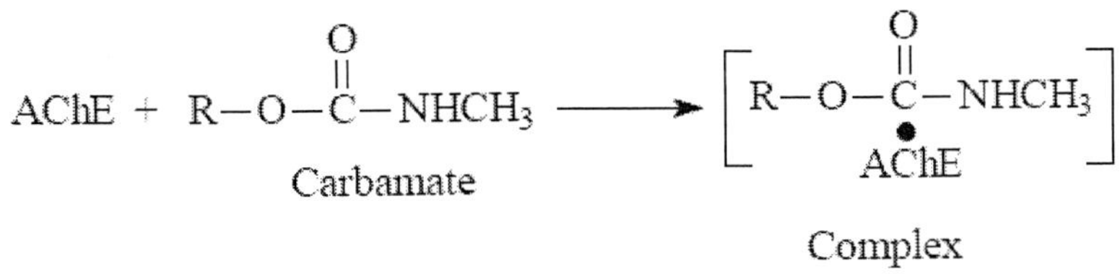

Unlike organophosphate pesticides (OPhP) carbamates interact with acetylcholinesterase within 30 minutes only, while OPhP – for 3 days. As a result, symptoms of carbamate poisoning can be seen in 5-10 minutes already, while OPhP poisoning symptoms are evident after passing much more time. The only exception is dichlorvos affecting typhoid flies one minute after application. And if in 30 minutes after carbamate treatment fly behavior is the same, then it will hardly change afterwards. Carbamate toxicity for mammals varies greatly (Table 9).

Table 9. Indicators of LD_{50} (mg/kg) toxicity for carbamate insecticides

Compound	Rats (orally*)	Rabbits (skin)
Pirimicarb	147	>500 (rat)
Carbaryl	246	-
Isoprocarb	485	>500 (rat)
Propoxur	95	>500
Propoxur	-	>1000 (rat)
Carbofuran	11	3400
Aldicarb	0.9	>5.0
Methomyl	17	5.0

Carbamate metabolism in nature is studied by identifying fragments of decomposition of radiolabeled ^{13}C insecticide molecules. Example of chlorpropham biodegradation is given below. Main stage is ester hydrolysis to unstable acids, alcohols and aromatic amines. Alkaline hydrolysis results in 3-aminophenol formation.

* Orally – denoting a drug to be taken by mouth

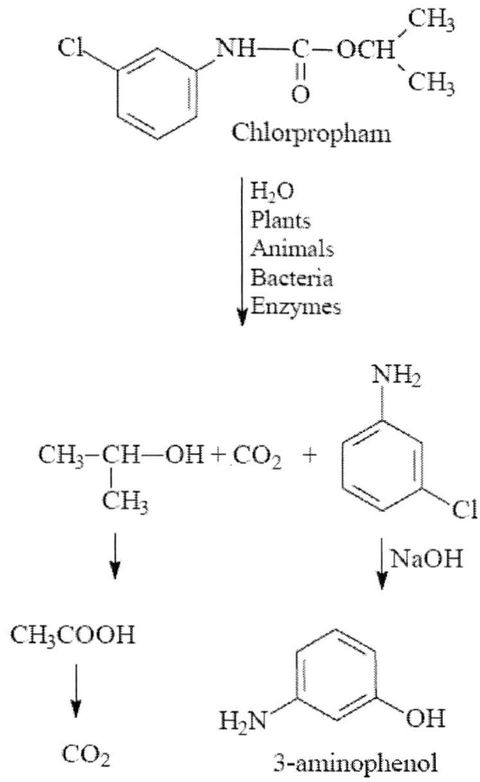
CH3
Cl
NH—C—OCH
O
CH3
Chlorpropham
H2O
Plants
Animals
Bacteria
Enzymes
NH2
CH3–CH—OH + CO2 +
CH3
Cl
NaOH
CH3COOH
H2N
OH
CO2
3-aminophenol

Chapter 10

10. Pyrethroids

Plant preparations have been used by people for insect control long ago. Since ancient times Dalmatian pyrethrum powder has been successfully used for parasitic insect control. Dalmatian pyrethrum does not occur in Russia. In nature it grows in mountainous parts of South Europe, mainly in Dalmatia (the Adriatic Sea shore), at 500-2000 above sea level (Figure 50).

Figure 50. Dalmatian pyrethrum.

Dalmatian pyrethrum reactants are esters of (1*R*)-*trans*-chrysanthemum and pyrethrum acids: pyrethrin I, pyrethrin II, cinerin I, cinerin II, jasmolin I and jasmolin II.

Insect susceptibility to pyrethrum deals with cuticle permeability. Pyrethrins and its analogues also affect insect nervous system, invoking shock in them (knockdown effect).

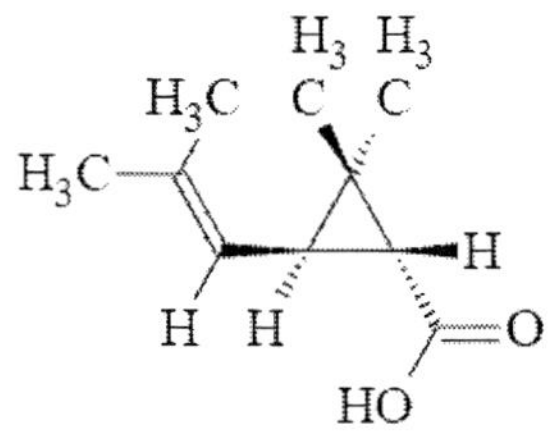

trans-chrysanthemum acid

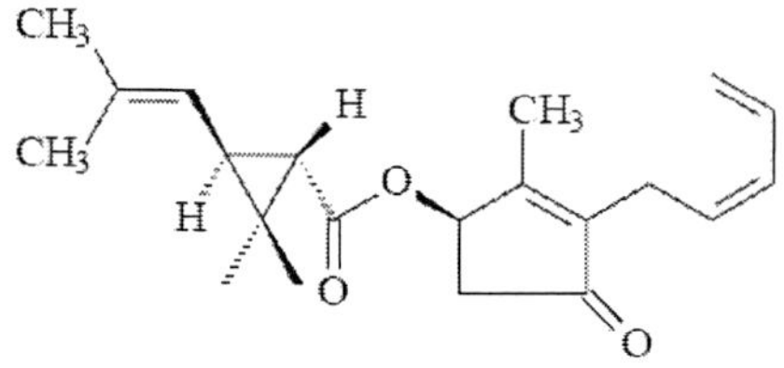

Pyrethrin I

Vegetable terpenoids such as carvone, Δ^3-carene, limonene, and others are the most promising original substrates for pyrethroid synthesis based upon them, as their structure contains necessary cyclopropane fragment with the required configuration of asymmetric carbon. Example of multistage synthesis of (1R)-*trans*-chrysanthemum acid – the key intermediate in pyrethroid synthesis based on selective transformations of (-)-carvone is given below [68]:

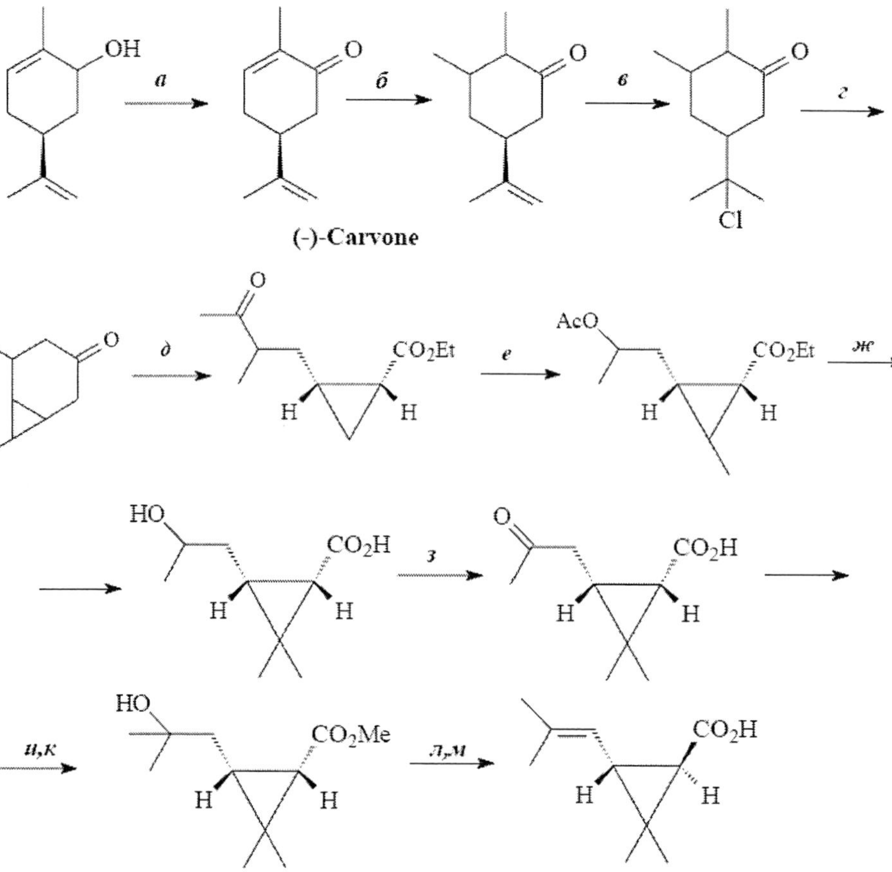

(-)-Carvone

(1R)-*trans*-chrysanthemum acid

Reagents: *а*.CrO_3; *б*.MeMgI/CuCl; *в*.HCl; *г*.NaOH; *д*.[O]; *е*.AcOOH; *ж*.NaOH; *з*.CrO_3; *и*.CH_2N_2; *к*.MeMgI; *л*.$POCl_3$; *м*.KOH

Schemes of synthesis considered above and similar ones, however, were not widely applied due to low output of desired product. More perspective is getting of acidic component by oxidation of (+)-3-carene with potassium permanganate. Together with splitting of double bond in (+)-3-carene, this reagent also regioselectively oxidizes carbon in allylic position to this double bond [69], which significantly decreases number of steps in insecticide synthesis.

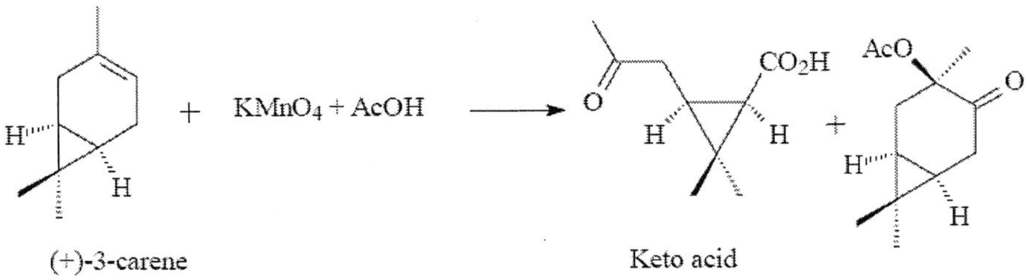

Keto acid like cyclic lactol - 4α-hydroxy-3-oxa-6,6-dimethylbicyclo[3.1.0]hexane-2-one (15) – can be obtained from (+)-3-carene applying ozonolysis in two steps [70].

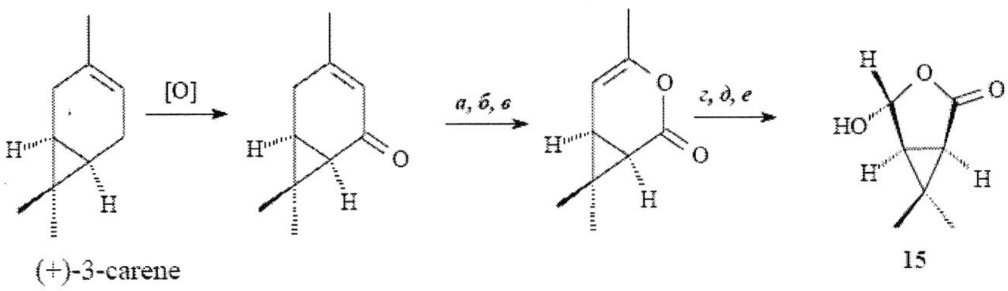

Low photostability of natural pyrethrins makes their field application ineffective. To find pyrethrins resistant to solar radiation numerous analogues were synthesized characterized by extremely high activity in relation to destructive insects under simultaneous resistance to destruction influenced both by ultraviolet irradiation and other environmental factors. As optically active forms of pyrethroids reveal more activity, efforts of scientists engaged in synthesis were devoted to elaboration of stereospecific methods for synthesizing new highly effective chiral• representatives. Pyrethroids belong to new generation preparations, with their world market share exceeding 25% (as compared with 1% twenty years ago).

• Chirality – a property of a molecule to be asymmetric being reflected in perfectly flat mirror and is necessary condition for optical activity of molecules.

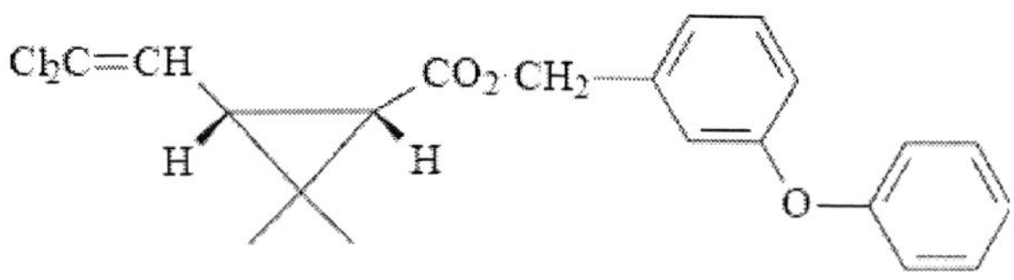

Ambush, permethrin, talcord, ecthyiban
LD_{50} = 400-4000 microgram/kg
Insecticide for granary protection

Great majority of photostable pyrethroids are derivatives of *meta*- phenoxytoluene (17), which can be easily obtained by Ullman reaction by substitution of halogen atom in aromatic ring to phenoxy group influenced by phenolate 16 over catalysts. Oxidation of *meta*-phenoxytoluene 17 to 3- phenoxyacetic acid is made by oxygen over catalysts, further hydrogen reduction under higher pressure or by alkali metal hydride leads to required alcohol 18 [71].

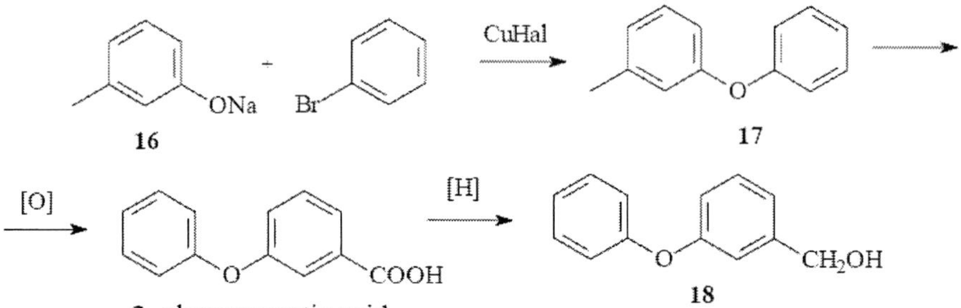

16 17 3- phenoxyacetic acid 18

Acidic component 21 can be obtained by different methods, e.g. by cyclopropanation of 2.5-dimethylhexa-2.4-diene (19) with diazo-acetic ester 20 over rhodium catalyst.

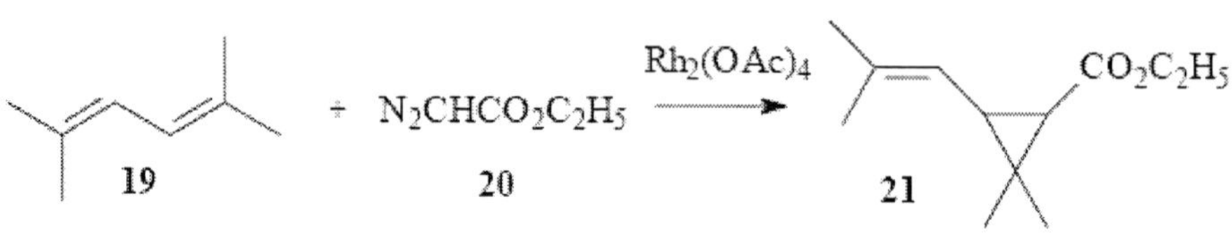

19 20 21

Sulphur yields 22 proved to be very good cyclopropanation agents.

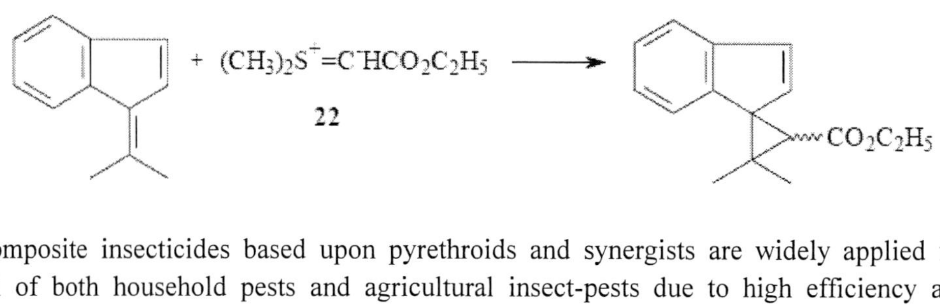

22

Composite insecticides based upon pyrethroids and synergists are widely applied for control of both household pests and agricultural insect-pests due to high efficiency and possibility to greatly decrease costs for expensive pyrethroids. Widely spread pyrethroid synergists – piperonyl butoxide (PPB) is bought abroad [72].

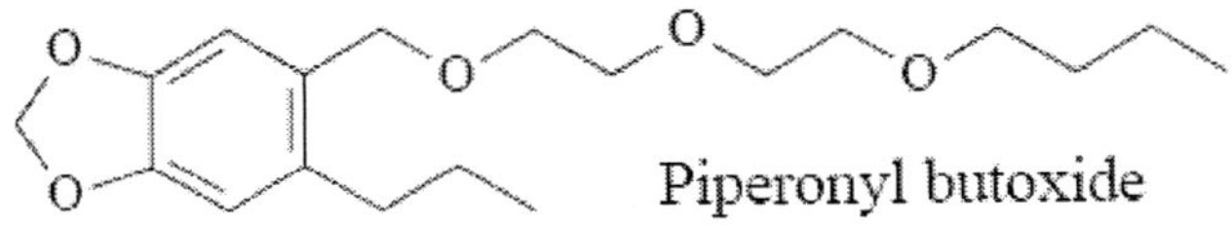

Piperonyl butoxide

A number of glycine derivatives O.O-dialkyl-S-[(N-acyl-N- alkoxycarbonyl methyl) aminomethyl]dithiophosphates and corresponding monothiophosphates (activation metabolite) with different alkyl substituents in phosphorus part of a molecule, in aminoacid and carbamate (for N-alkoxycarbonyl derivatives) ester groups and different N-acyl groups, as well as a number of derivatives of β-alanine - O,O-diethyl-S-[(N- alkoxycarbonyl -N-2-alkoxycarbonylethyl)aminomethyl] dithi- and thiophosphates have been examined for toxicity on two sensitive and two resistant races of typhoid flies and three species of synanthropic cockroaches (red, black and ash-grey). Their antiestarase activity as well as synergetic activity of compounds in relation to permethrin of these insects have been studied. While examining derivatives of 1,2,3-oxazaphosphorinane different contribution of monooxigenase (MO) and carboxylesterase (CE) to pyrethroid detoxification in fly and cockroach organisms is defined. Thereby "universal" synergist is to suppress simultaneously actions of these two enzymes. Testing of composite preparations based upon permethrin and two mostly active compounds (by one for each group) regarding Colorado beetle and melon aphid has confirmed high efficiency of tested synergists, allowing 10-30 times decreasing costs for expensive pyrethroid.

As a rule, insecticide synergists are enzyme inhibitors either participating in insecticide detoxification, or playing vital functions in insect organisms. For example, enzymes of respiratory chain are activated by one of well-known synergists of pyrethroids and other insecticides - DEF (diisopropyl fluorophosphate).

Chapter 11

11. Nicotine and Anabasine

Nicotine (after a French name J.Nicot who was the first to bring in tobacco to France in 1560) is an alkaloid contained chiefly in leaves and seeds of different sorts of tobacco-plants (*tabaco* - Spanish). This liquid with bad smell and pungent taste was isolated in pure form in 1828 by Posselt and Reimann.

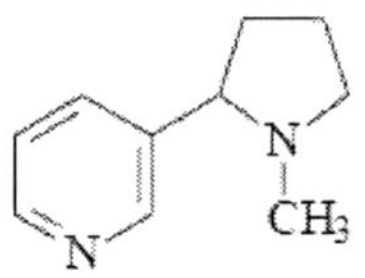

Nicotine

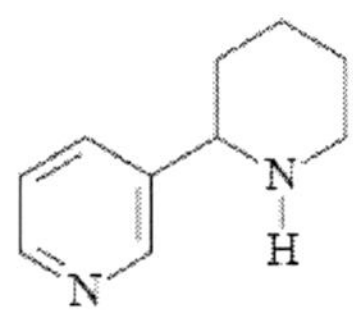

Anabasine

Over 70 cultures of wild tobacco plants are known to be found in the tropics, the USA and Australia. Two sorts of smoking tobacco are cultivated (not occurring in nature): rustic tobacco (*Nicotiana rustica*) (Figure 51) and common tobacco (*Nicotiana tabacum*). Annual production of leaf tobacco exceeds 5 mln tons.

Figure 52. Nicotiana rustica.

Nicotine is a very toxic substance successfully applying as insecticide.

Another alkaloid is anabasine, structural analogue of nicotine and also used as insecticide is isolated from aphyllous anabasis (*Anabasis aphylla*) – succulent suffrutice of goosefoot.

Anabasis grows as flat globe-shaped bush of 55-60 cm (up to 140 cm) in diameter. The leaves contain no chlorophyll: assimilation is made by one-year stems. It grows in even lands of Kazakhstan, in republics of Central Asia, Azerbaijan and in south-eastern parts of European part of Russia. Despite high contact insecticidal activity both nicotine and anabasine are not widely applied due to high toxicity for warm-blooded animals.

At present, neonicotinoids are of greater practical interest. They belong to pyridine derivatives, e.g. imidacloprid – systemic insecticide applied for seed treatment is 1-[(6-chlor-3-pyridinil)methyl]-*N*-nitro-2-Imidazolidimine.

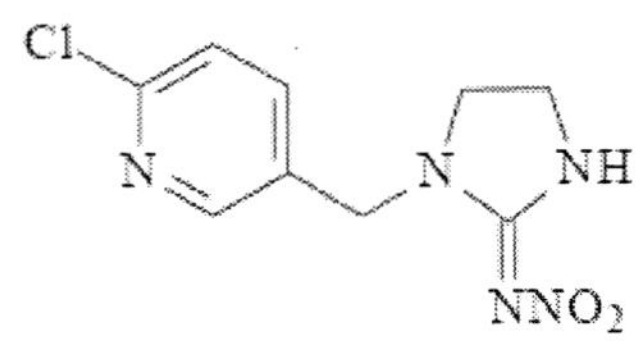

Imidacloprid

This is one of the most active insecticides similar by its bioactivity to pyrethroids, and even exceeding OPhC and carbamates. Imidacloprid suppresses activity of acetylcholinesterase, and it is antagonist of nicotine-acetylcholine postsynaptic receptors and induces hyperpolarization of nerve fiber membranes in insects. Certain structural similarity and similar toxic action mechanism made it possible to refer this insecticide to nicotine analogs i.e. neonicotinoids [73].

Imidacloprid is synthesized by interaction between nitroguanidine 23 and ethylene diamine 24 with further alkilation of imidazolidine-2-nitroimin (25) 3- chloromethyl-6-chloropyridine (26):

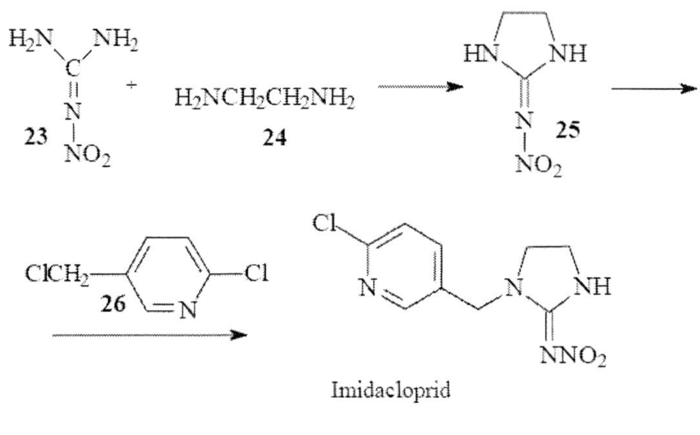

Imidacloprid

- In small doses hydrochloride anabasine is recommended as a drug for smoking avert. For this purpose it is produced as pills, films and “Gamibasinum” chewing gum.

Chapter 12

12. Rotenone and Its Analogs

Water extracts from some plant species *Derris elliptica* (Figure 52), *Lonchocarpus spp. Thephrosia spp* (*Leguminosae*) and related species were applied in Asia, Africa and Indonesia for fishing, as 5-8% concentration of rotenone in them is quite toxic for fish. Powder from these plants wood was also used against household ectoparasites. Agent as white crystalline powder insoluble in water, but soluble in most of organic solvents was extracted by Geoffrey [74] and Nagai [75].

Figure 52. Derris elliptica.

Its structure was defined by USA researchers in 1933 and the substance was described as 1,2,12a-tetrahydro-8,9-dimethoxy-2(1-menthylethenyl-(1)benzo-pyrano(2,4-b)furo(2,3-h)(1)benzofuran-6(6H)-one (rotenon). Rotenon synthesis and biosynthesis are described by

Crombie [76] however it is obtained by extraction from plants, because it proves to be economically more practical. Rotenon toxicity for the cold-blooded is stipulated by affecting the process of breathing within cell mitochondria.

Insecticide activity of the substance isolated from plant *Ryania speciosa* was originally describe in 1945. It was shown that the agent was ryanodine. It interacts with calcium channels of sarcoplasmic reticulum resulting in calcium accumulation lethal for cells and muscle paralysis. LD_{50} of ryanodine for rats is 1.2 mg/kg. When rats and cavies are fed with food containing 1% of ryanodine for 6 months, no visible changes take place in their organisms. Large doses of toxin (approx. 1500 mg/kg per day) in rat food caused hemorrhage in pancreatic gland and intestine. The preparation is stable enough at storage and under light. Biodegradation results in anhydrooxyryanodine. Ryania preparations have been originally produced by *S.B. Penick and Co*, but they were not widely applied.

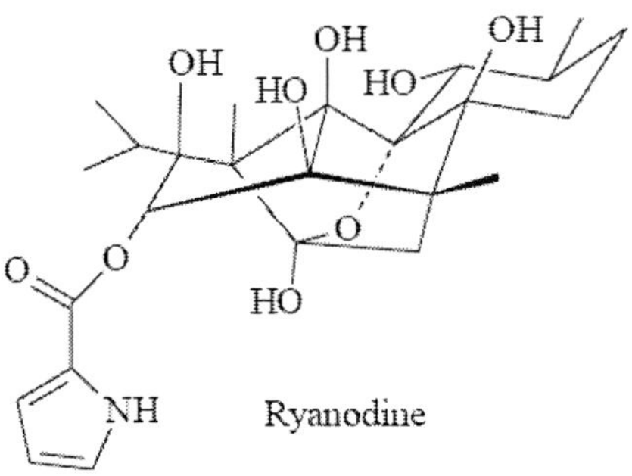

Ryanodine

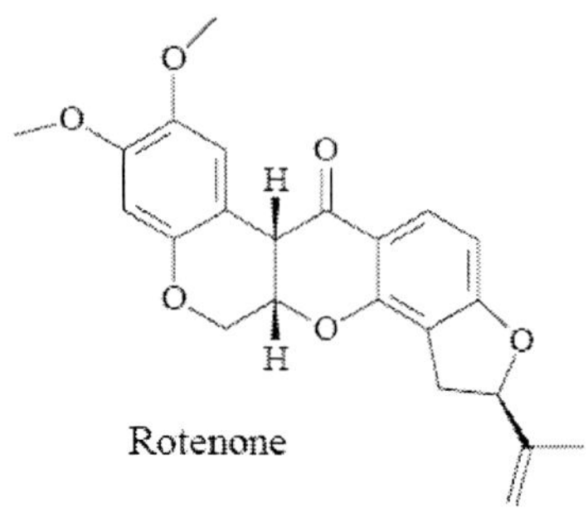

Rotenone

Whole synthesis of ryanodine and ryanodol was made by Belanger in 1979 [77], vanillin being obtained by extraction from *Ryania speciosa*..

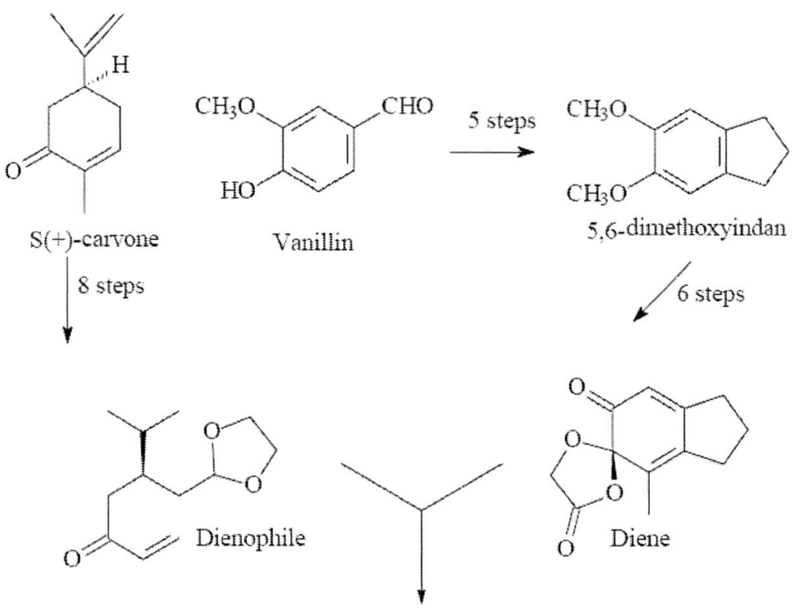

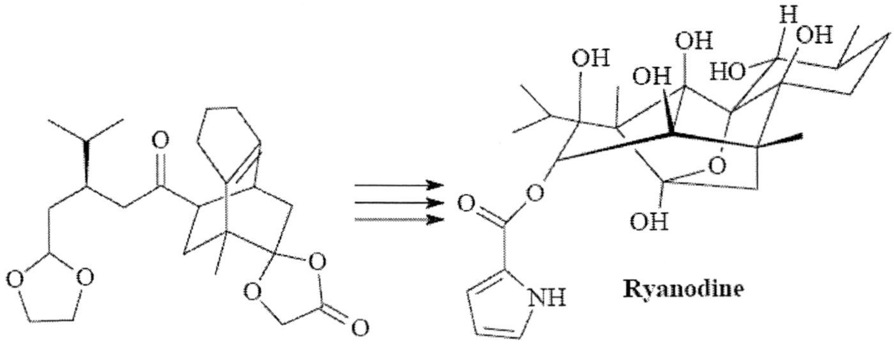

Ryanodine

Chapter 13

13. Nereistoxins

Observing flies feeding on decomposing marine worms *Lumbriconereis heteropoda* and dying immediately afterwards made it possible to define that these worms contain toxin lethal for insects. Isolated and identified new insecticide was N,N-dimethyl-1,2-dithiolane-4-amin and was named as nereistoxin.

Nereistoxin

Later this natural product together with great number of its analogs have been synthesized and tested for insecticidal activity. It has been shown that only those substances which transform in insect organism into native* nereistoxin are active. Two such synthetic analogs have been made for control of Coleoptera and Lepidoptera insects. Cartap revealed insecticidal properties of wide spectrum of activities against pests of rice stems, while bensultap was used for Colorado beetle and other pests control. Sandoz company has elaborated analog of nereistoxin – thiocyclam for control of wide spectrum of Coleoptera and Lepidoptera pests of some cereals.

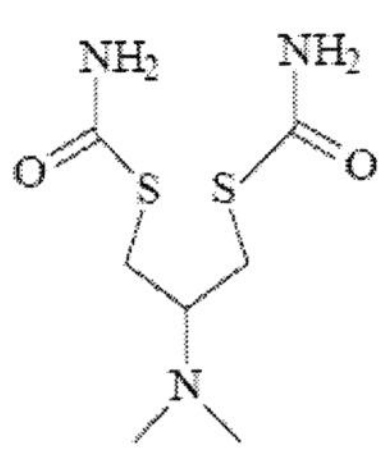

Cartap

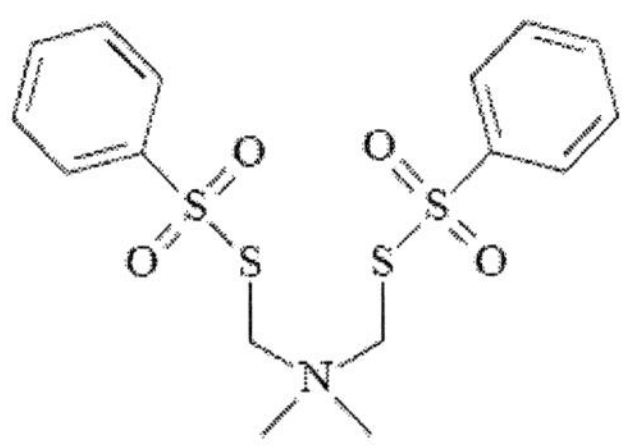

Bensultap

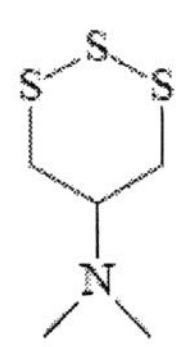

Thiocyclem

* Native – from English *native*, meaning substances found naturally and preserving their biological activity.

Thiocyclam can be obtained under the following scheme [78]:

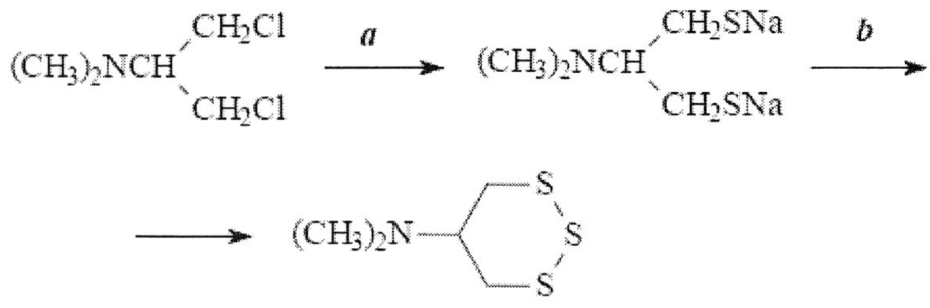

Reagents:: a.Na_2S; б.SCl_2

Cartap hydrochloride synthesized originally in 1967 in Japan was widely applied as soluble powders, dusts and granules for control of harmful insects in Asia, Europe and North America. It is acid-resistant, but in neutral or alkaline medium it is hydrolyzed to dihydronereistoxin which is easily oxidized in air to nereistoxin:

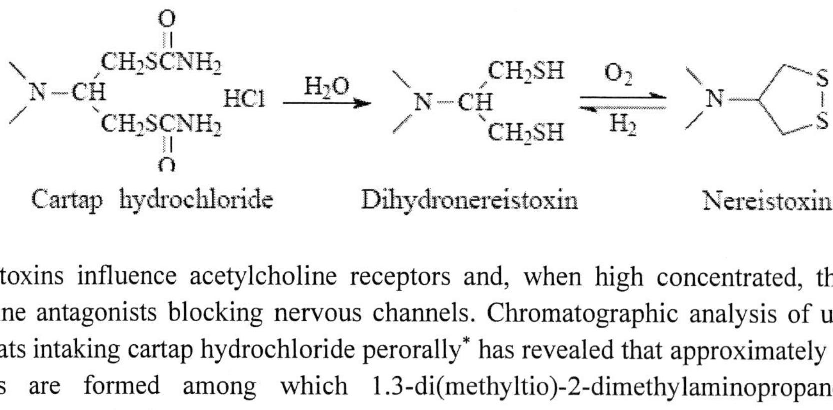

Cartap hydrochloride Dihydronereistoxin Nereistoxin

Nereistoxins influence acetylcholine receptors and, when high concentrated, they are acetylcholine antagonists blocking nervous channels. Chromatographic analysis of urine of mice and rats intaking cartap hydrochloride perorally* has revealed that approximately 9 main metabolites are formed among which 1.3-di(methyltio)-2-dimethylaminopropane was identified. No cartap hydrochloride has been discovered in the urine.

Preparation containing ^{35}S isotope was synthesized for studying cartap migration within plants. Rice plants reveal that maximum concentration of the agent is accumulated in plants on the 7th day, and residual radioactivity is seen within 30 days after plant treatment. Intravenous introduction of cysteine to rabbits after acute poisoning with cartap (LD_{50} 1120-1150 mg/kg) eliminates symptoms of poisoning animals.

2-aryl-5-dimethylamino-2-cyan-1.3-dithiazine (28) with high insecticidal activity is synthesized by interaction of 2-dimethylamino-1.3-bis(arylsulfontio)propane (27) with CH-acids. At microsomal oxidation influenced by enzymes it is transformed into nereistoxin and its S-oxide, thereby explaining insecticidal activity of this compound [79].

* Perorally – introduction into animal stomach through the mouth.

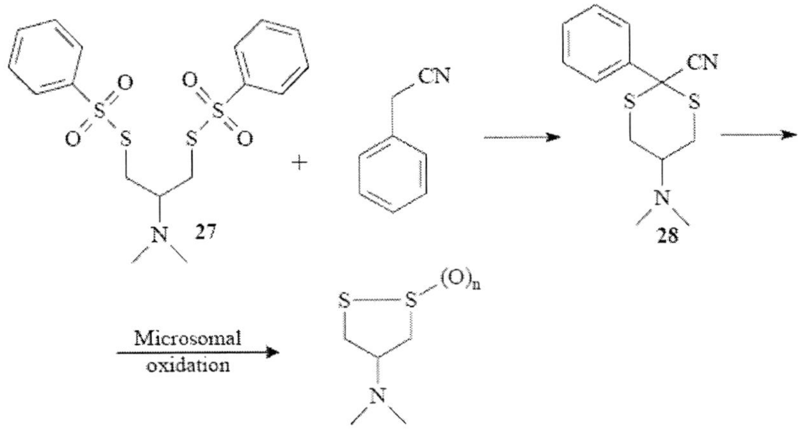

n = 0 (nereistoxin) 1(S-oxide)

Guinesins A, B and C taking quite strong insecticidal activity [79] and being structural analogs of nereistoxin were isolated from the brazil-wood *Cassiopourea guianensis* by the end of 80s:

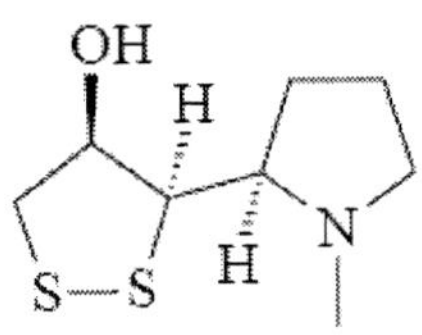

Guinesin A

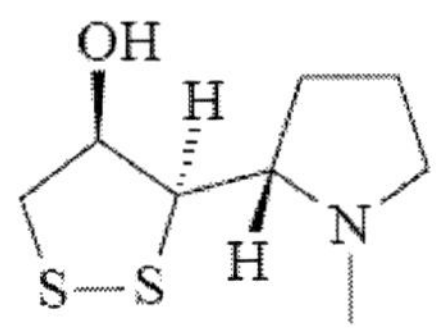

Guinesin B

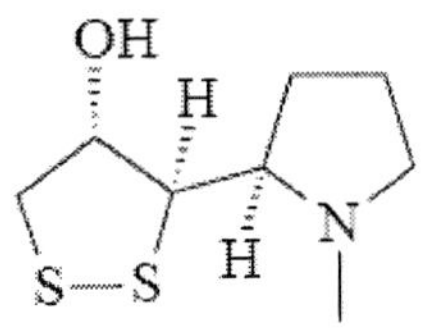

Guinesin C

While modifying guinesin molecules a number of dialkylaminomethyl-1.2-dithiolanes, 1.2.3-trithianes and 2.2-disubstituted of 4- dialkylaminomethyl-1.3-dithians was received. By their activity 3-dimethylaminomethyl-1.2-dithiolane and *cis*-3.5-bis(dimethylaminomethyl)-1,2-dithiolane are similar to nereistoxin:

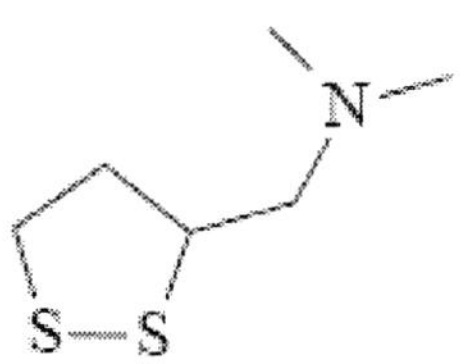

3-dimethylaminomethyl-1.2-dithiolane

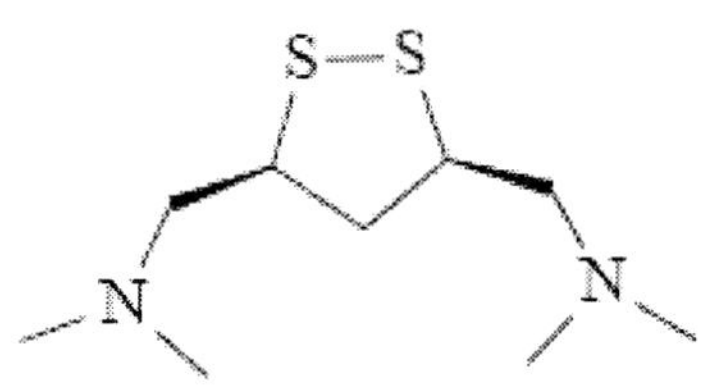

cis-3.5-bis(dimethylaminomethyl)-1,2-dithiolane

Chapter 14

14. Juvenile Hormones of Insects and Their Analogs

14.1. Biological Effect and Synthesis

At the end of 30s of the last century it was discovered that substances of terpenoid structure named as juvenile hormones (JH) actively participated in regulating insect development on larval stage. By now five native juvenile hormones have been isolated and identified: JH-1, JH-II, JH-III, JH-0 and 4-Me-JH [81].

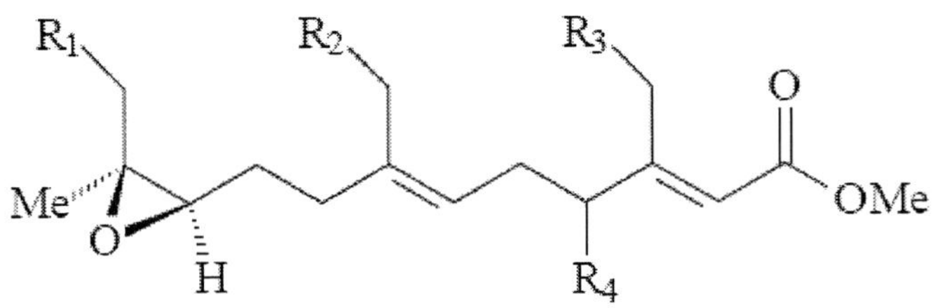

JH-I (R_1=R_2=Me, R_3=R_4=H); JH -II (R_1=Me, R_2=R_3=R_4=H);
JH-III (R_1=R_2=R_3=R_4=H); JH -0 (R_1=R_2=R_3=Me, R_4=H);
4-Me- JH -I (R_1=R_2=R_4=Me, R_3=H)

Observation for insects has revealed role of hormones in regulating metamorphosis (Figure 53), being shown as realized by some hormones. At each developmental stage accompanied by metamorphosis neurosecretory cells of insect brain produce brain hormones, stimulating synthesis of steroid hormone in the prothoracic gland to secrete molt – ecdysone. Juvenile hormone is produced by two small special glands (*corpora allata*) to be found in insect head. This hormone does not suppress ecdysone activity and provides for transfer from molt to the next larval stage. At the last larval stage juvenile hormone is not produced, so next molt leads to pupation and transfer to the next larval stage in insect development. In Samia cecropia prothoracic gland, stimulated by brain hormone starts to secrete ecdysone – molt-initiating hormone, which larva comes through during metamorphosis (See Figure 53).

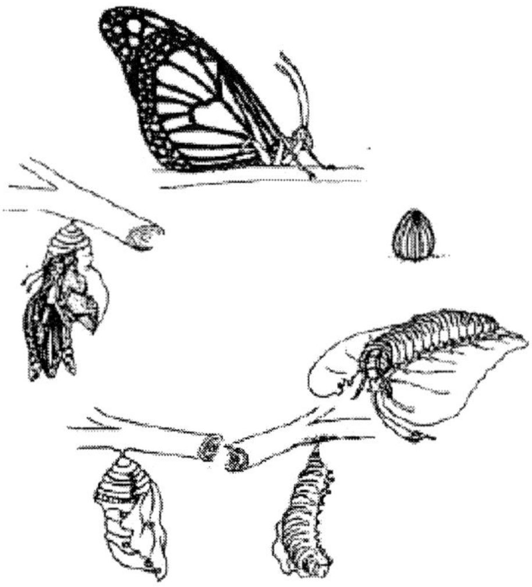

Figure 53. Metamorphosis.

Thus, under normal insect development at metamorphosis of larva into imago JH concentration in insect organism is equal or similar to zero. Introduction of JH into insect organism at this developmental stage tragically affect the whole insect population due to disruption of its normal progress, resulting in generation cut. Considering such opportunity to regulate insect population, JH are regarded to be environmentally safe and selective insecticides.

As amount of JH in insect organisms is too small, it is impossible to extract them from natural sources for practical application, and the only real way for their obtaining is chemical synthesis. Perspectives for practical application of these insect growth regulators have been found after discovering JH effect in some natural substances and the whole range of synthetic compounds structurally similar to native JH. These juvenile hormone analogs (JHA) were named juvenoids. Compounds with far more JH-activity than natural JH have been found among great number of JHA. Active juvenoids are found not only among substances similar to native JH, but in compounds which composition is remotely similar to JH structure.

In extracts of *Ageratum houstonianum* (Figure 54) Bowers W.C. discovered two compounds which affecting *Oncopeltus* and *Dysdercus* bugs result in skipping nymphal stage by immature species and transforming into tiny female or male sterile species unable for reproduction. Substances contained in extracts of this plant were identified as 7-methoxi- and 6.7-dimethoxi-2.2-dimethylchromens and named precocenes. Actually the are JH antagonists. Their wide application is limited by their high cost and difficulties in using as these preparations are to be thoroughly measured [82].

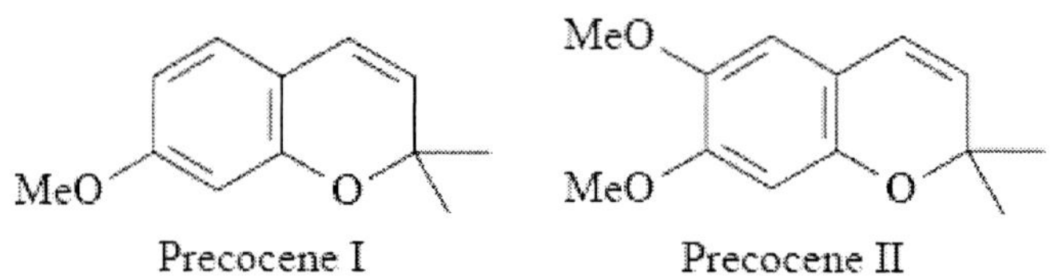

Another precocene – juvoocimene II has been extracted from sweet basil.

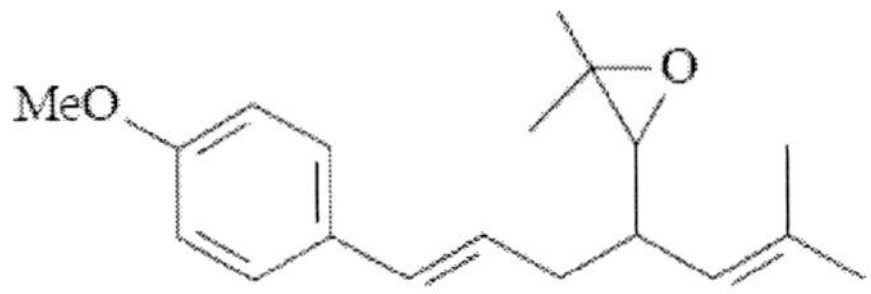

Juvoocimene II

Figure 54. Ageratum houstonianum.

- One of species of animalcular parasites *Nosema* (*Protozoa*) infecting flour beetle larvae synthesizes similar analog of flour beetle juvenile hormone. Millions of these parasites swarm starting to produce juvenile hormone in the host larvae preventing the latter from transformation into adult insect. Instead larva continues to grow reaching gigantic sizes twice exceeding its normal adult size by weight, and represents the so-called biocombine for development and reproduction of these parasites [83].

14.2. Synthesis of Natural Juvenile Hormones

Stepwise mounting method is usually used for synthesis of long-chain molecules applying practically suitable reactions. If double bond is to be introduced, Wittig reaction is applied. Phosphonate modification of Wittig reaction usually used for mounting trisubstituted double bond leads to mixture of geometrical isomers. Thus, literature describes synthesis of JH-I as mixture of (Z) and (*E*)-isomers under Δ^2- and Δ^6-double bonds by three-stage condensation of thrimethylphosphonacetate with methylethylketone.

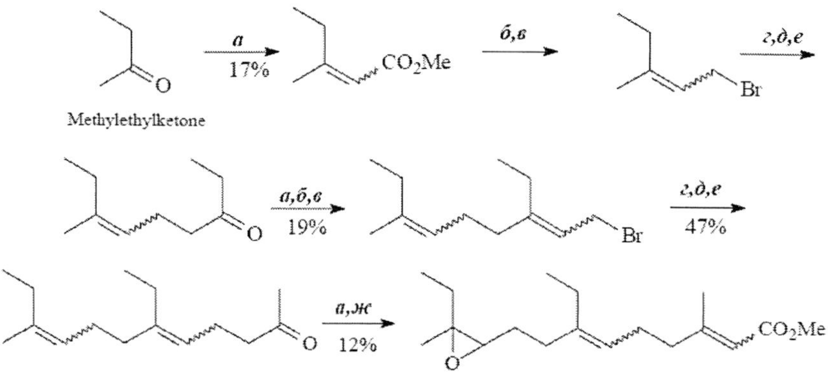

JH-1

Reagents: *а.* $(MeO)_2P(O)\overset{\ominus}{C}HCO_2Me$; *б.*$LiAlH_4$; *в.*$PBr_3$; *г.*Mg; *д.*$MeCH_2COCH_2CO_2Et$; *е.* heating; *ж.* m-$ClC_6H_4CO_3H$

However, this multi-stage method is not very effective (some yields reach only 12-17%). Even optimization and increase of yields of reaction products on some stages cannot be considered effective, because total yield of end product becomes low anyway as a result.

From this point of view method applying large-block mounting of JH molecule from two fragments proves to be more successful.

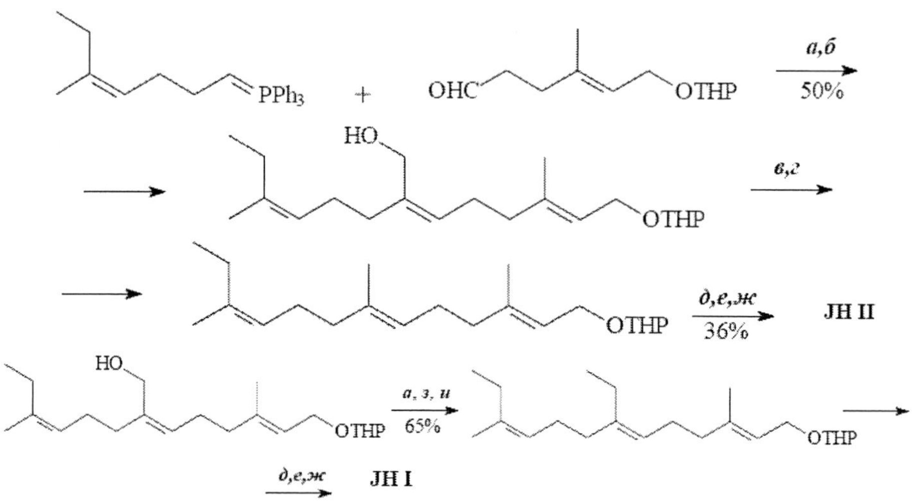

Reagents: *а.* n-BuLi; *б.*CH_2O; *в.*TsCl/Py; *г.*$LiAlH_4$ *д.*MnO_2; *е.*MnO_2/NaCN/MeOH; *ж.*m-$ClC_6H_4CO_3H$; *з.*CH_2=PPh_3; *и.*H_2O_2/EtOH/NH_2NH_2

Application of cyclic trimer of isoprene (29) as initial compound for key synthon (30) in synthesis of JH-III allows a number of stages to be significantly decreased, thereby increasing total yield of end product. The key stage in this synthesis is partial ozonolitic disintegration of one of trisubstituted bonds in 1,5,9-trimethyl-1,5,9-cyclododecatriene (29). Ketoacetal (30) resulting after treating reaction mixture with salmiac in methanol is a universal synthon for both JH and their analogs.

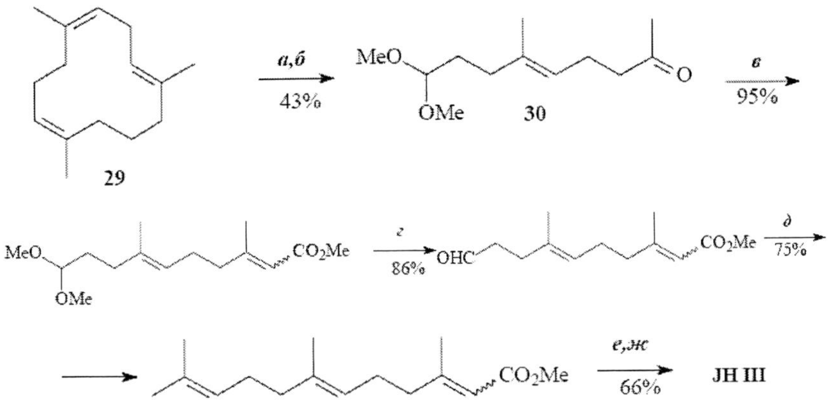

Reagents: *а*.O_3/MeOH; *б*.MeOH/NH_4Cl; *в*.$Me_3SiCH_2CO_2Me$/iso-Pr_2NH; *г*.Py TsOH/Me_2CO; *д*.$Me_2C{=}PPh_3$; *е*.NBS/THF; *ж*.K_2CO_3/MeOH

Individual (2*E*.6*E*)-isomers which are more active, as a rule, than other isomers can be selected by column chromatography.

Stereospecific syntheses of JH-II are made from terpenes already containing major fragment of JH molecule: (*E,E*)-farnesol, methyl-(*E*)-geranoate and (*E,E*)-farnesoate. Main sources of terpens are aspic oils of spice-aromatic plants and their gums.

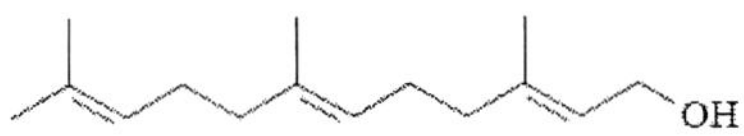

(E,E)-farnesol

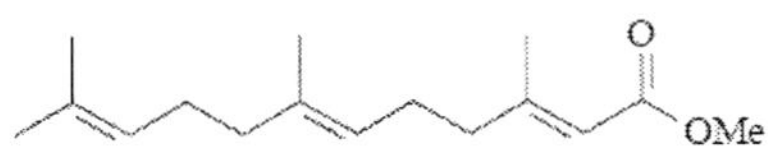

Methyl-(E,E)-farnesoate

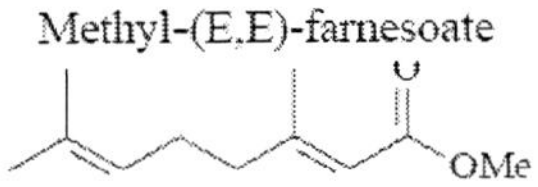

Methyl-(*E*)-geranoate

Epoxidation of terminal double bond in methylfarnesoate (31) by Van-Temelen resulted in JH-III. Further opening of epoxide to diol (32) and splitting of the latter with sodium periodate resulted in aldehyde (33) to be transformed later into specific JH-II.

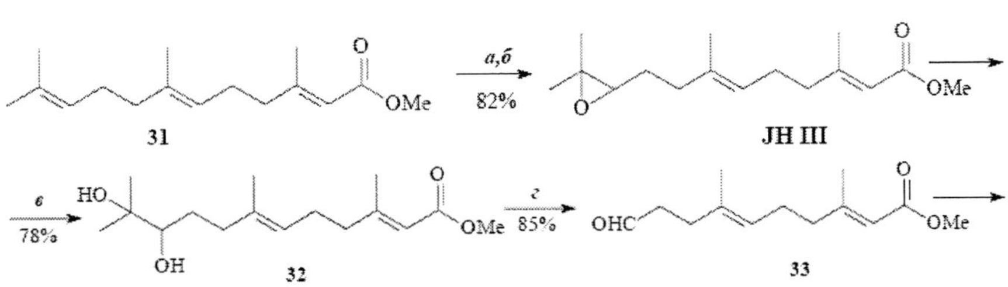

Distillation of JH-III leads to allyl alcohol (34) with high yield, which acetate (35) was alkylated by dimethyllithiumcuprate forming JH-II:

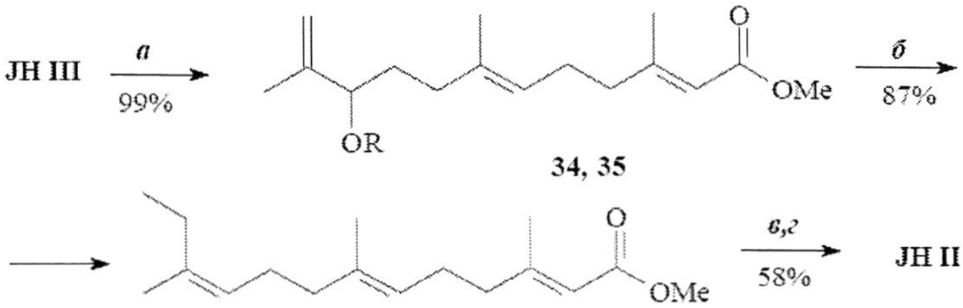

34, 35

R=H (34), Ac(35)

Reagents: *a*. Heating; *б*.Me_2CuLi; *в*. NBS; *г*.K_2CO_3/MeOH.

Still in 1970 it was discovered that optically native JH have (10*R*,11*S*)-configuration. Futher investigation revealed that synthetic (+)-enantiomers corresponding to such configuration are more highly biologically active as compared with racemates. One of synthetic approaches to optically active JH is based on asymmetric epoxidation of allyl alcohols by Sharpless with tert-butyl hydroperoxide over isopropoxy titanium. Synthesis of (10*R*,11*S*)-JH-III containing tritium at C_{13}-atom, allowing metabolic transformations of JH within insect organism to be examined, is made by allyl oxidation of methylfarnesoate (36) into terminal position with further asymmetric epoxidation of formed allyl alcohol (37). Deoxygenation of tritiated oxy-epoxides through iodide leads to 13*T*-JH-III.

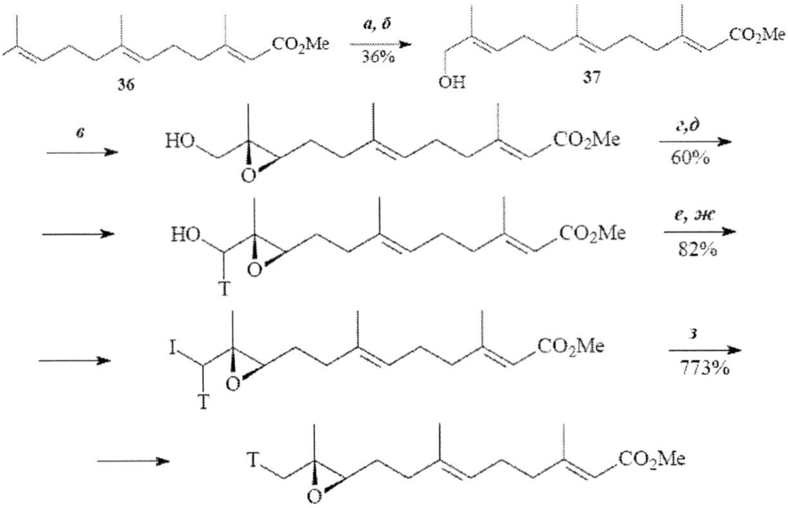

36 37

(13T)-**JH III**

Reagents: *a*. tret-BuOOH/SeO_2; *б*.$NaBH_4$;
в. (iso-PrO)$_4$Ti/ tret-BuOOH/iso-·$C_3H_7OOCCH(OH)CH(OH)COOC_3H_7$-iso;
г.CrO_3/Py; *д*.$NaBH_4$; *е*.TsCl/Py; *ж*.NaI/Me_2CO; *з*.$NaBH_3CN$

Asymmetric epoxidation also has served as the key stage in synthesis of optically active (10*R*,11*S*)-JH-I from homoallyl alcohol (38). Further transformations not affecting chiral centers have led to (10*R*,11*S*)-enantiomer similar to either natural JH, or its (12*D*,13*D*)-analog. The resulting products had high optical purity (95%). Wittig reaction is applied here for making trisubstituted double bond like in examples described earlier. Use of different protection for hydroxy groups allowing selective oxidation of one of hydroxy groups to carboxyl ones to be made is of interest in this synthesis.

- At high intensity of population the beetles – meal worms (*Tenebrio*) eat their eggs, thereby controlling population growth. When making complex experiments for corn grain storage where four pest species are available the Confused flour beetle (*Tribolium confusum*) is found to be dominant species decreasing population of all other insects [84].

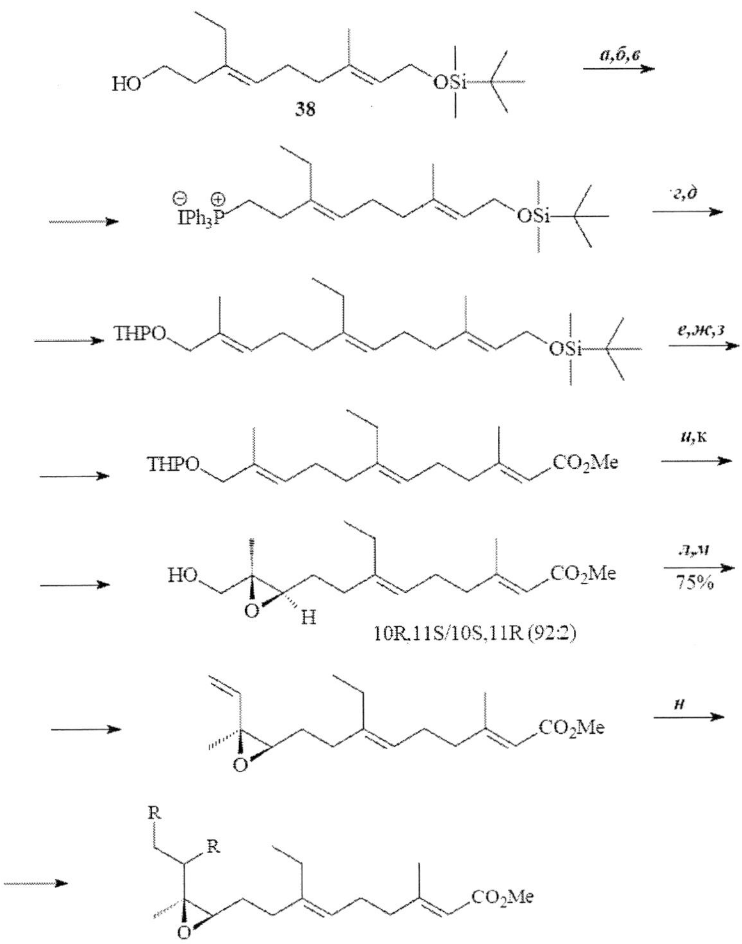

R=H(JH-I); D(12D, 13D-JH-I)

Reagents: *а*. TsCl/Py; *б*.NaI; *в*.Ph_3P; *г*.BuLi; *д*.MeC(O)CH_2OTHP; *е*.Bu_4NF; *ж*.MnO_2; *з*.MnO_2-NaCN-MeOH; *и*.TsOH Py MeOH; *к*. tret-BuOOH-CH_2Cl_2-Ti(OPr--iso)$_4$-$C_2H_5OOCCH(OH)CH(OH)COOC_2H_5$; *л*.$CrO_3$/Py; *м*.$Ph_3P{=}CH_2$; *н*. H_2 or D_2/Rh(Ph_3P)$_3$Cl

Kinetic separation is used in synthesis of active JH-III at interaction between racemic epoxide (39) with optically active organoaluminum reagents (40). Further transformations of chiral (*R*)-epoxide (41) resulted in (*R*)-(-)-JH-III:

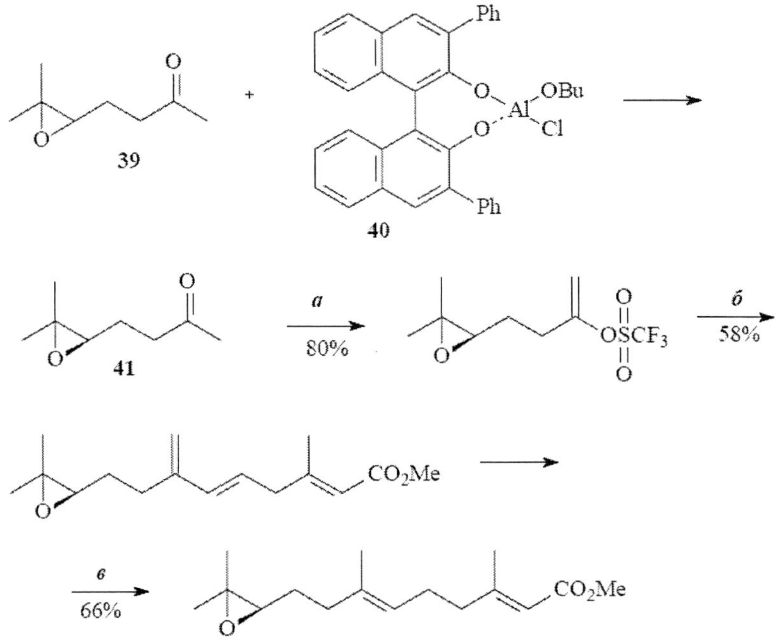

R-(-)-JH-III

Reagents: *a*.4- tret-$BuC_6H_4NSO_2CF_3$/LDA(lithiumdiethylamid); *б*.$Pd(PPh_3)_4$.

CsF(THF). Bu_3Sn CO_2Me; *в*.$H_2/Cr(CO)_3$ (80 atm)

Enzymatic reactions are successfully used in synthesis of optically pure JH. So, racemic 10,11-epoxy farnesol (42) influenced by fungus culture *Helminthosporium sativum* was transformed into (-)-*threo*-10,11-dihydrofarnesyl acid (43) and further to (10*R*)-JH-III. Similarly optically pure (10*R*,11*S*)-JH-III was obtained from homofarnesol.

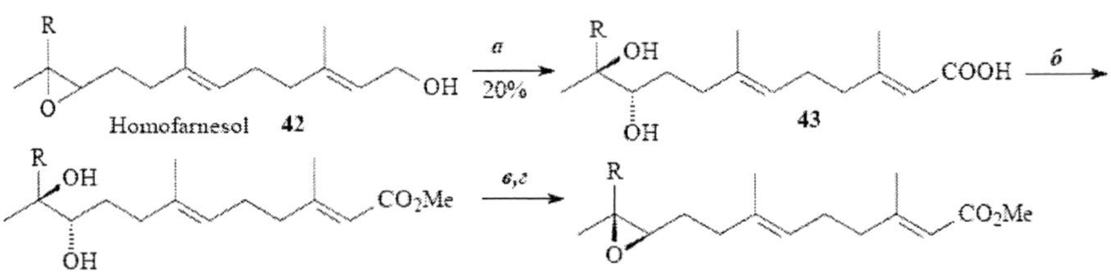

R=Me[(10R)- JH-III]; Et[(10R, 11S)-JH-II]
R= Me, Et

Reagents: *a*. *Helm. Sativum*; *б*. CH_2N_2; *в*.MsCl/Py; *г*. KOH/MeOH

14.3. Synthesis of Natural Analogs of Juvenile Hormones

Discovery of juvenoids in plants was not an occasional fact. Being on scientific trip in the USA Czech scientist Karel Slama cultivated a soil insect - red soldier bug *(Pyrrhocoris apterus L.)* on filter paper.

He was surprised having seen that insect metamorphosis had been disturbed with no pupation at the last larval stage. The matter was in the origin of the filter paper. In this case it was produced from balsam fir *(Abies balsamea)* (Figure 56). Metamorphosis continued smoothly with other sorts of paper.

Figure 55. Echinacea angustifolia.

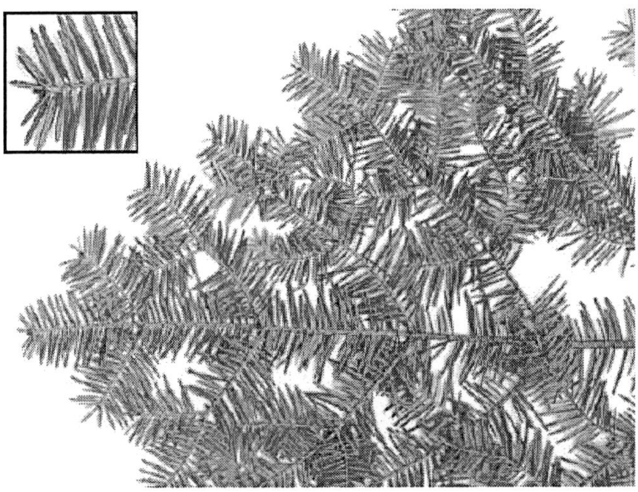

Figure 56. Abies balsamea.

Juvabione which is a structural analog of juvenile hormone selectively affecting this very insect has been isolated as a result of extraction. Structure of an active agent proved to be identical to methyl ether of (+)-todomate acid with (*R,R*)-configuration of asymmetric centers. The new analog was named juvabione. A substance with similar structure, but with (*R,S*)-configuration was found in some plants. It was named epijuvabione. Together with (+)-juvabione one more active analog dihydrojuvabione was isolated from *Slovak fir.*

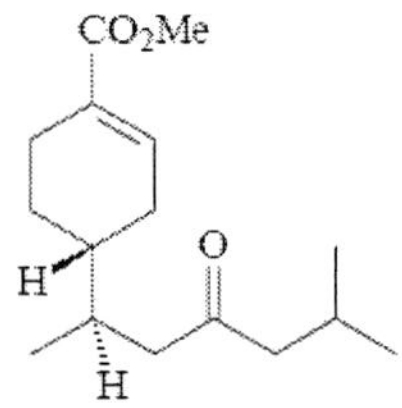

Juvabione

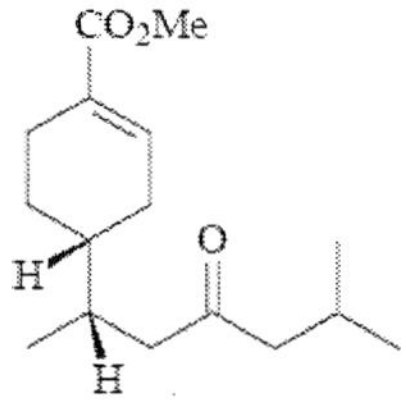

Epijuvabione

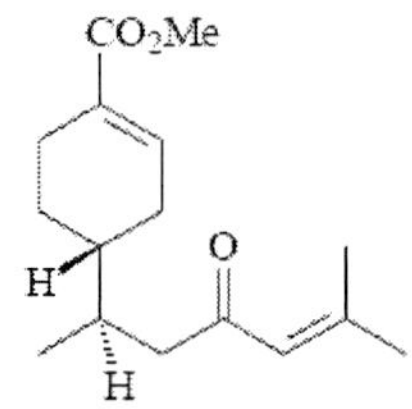

Dehydrojuvabione

Together with juvabione and epijuvabione the ar-juvabiones being their aromatic analogs possessing high JH-activity were isolated from natural objects. In 1975 optically active agent revealing juvenoid action in relation to yellow mealworm beetle, being a standard test-object for revealing JH-activity was isolated from coneflower (*Echinacea angustifolia*) (Figure 55). This substance named as echinolone was identified like (+)-(*E*)-10-hydroxy-4,10-dimethyl-4,11-dodecadiene-2-one.

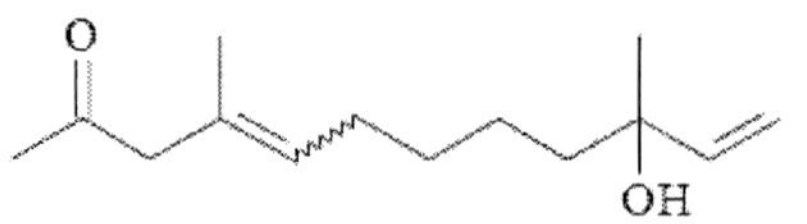

Echinolone

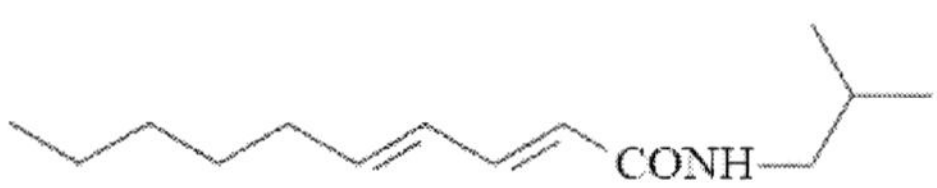

Pellitorine

N-isobutyl-2E,4E-decadienamide revealing rather high JH-activity and named as pellitorine [85] was isolated from *Anacyckus Pyrethrum roots, Piper nigrum* (Figure 57) and a number of other plants.

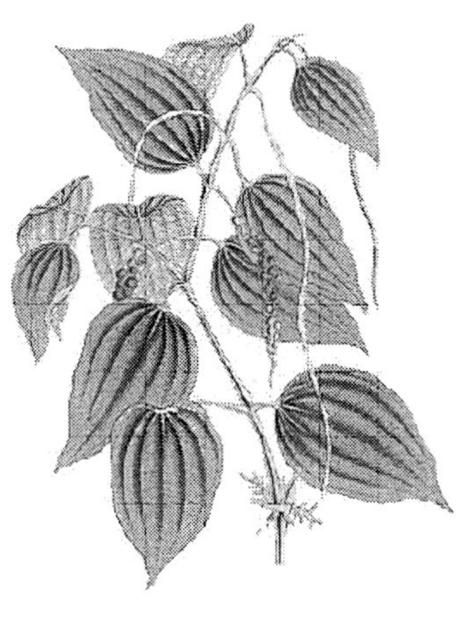

Figure 53. Piper nigrum.

Other three dienamides proving to be far more stable and highly active to insecticides were isolated from *Piper nigrum* by Sumitomo researchers [86].

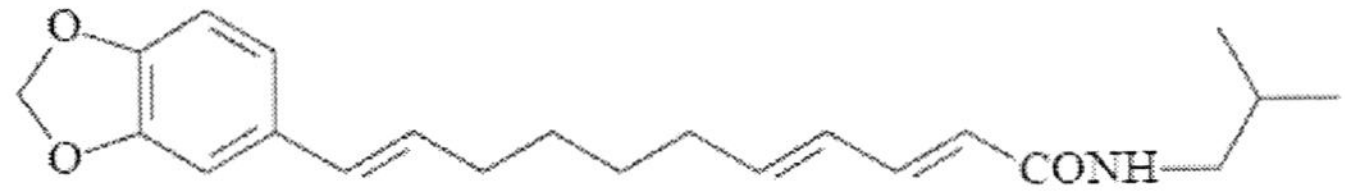

Pipercide

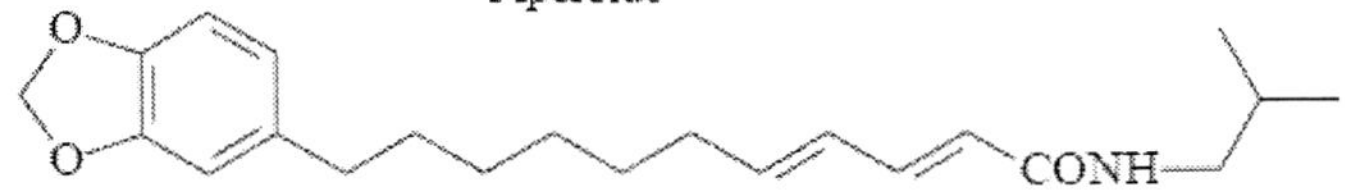

Dihydropipercide

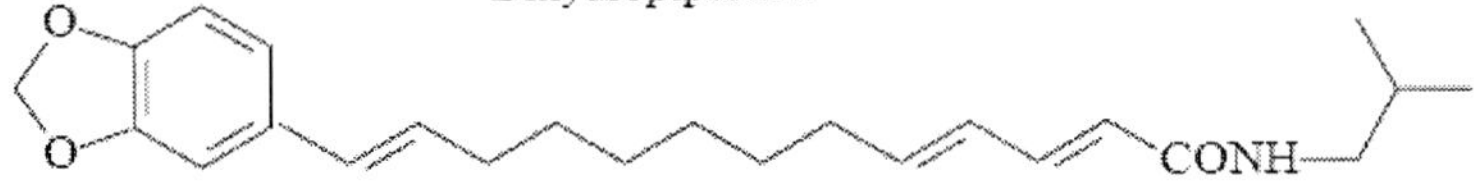

Quinensine

Among proposed non-stereospecific ways of synthesizing mixture of juvabione and epijuvabione the way based upon transformations of methyl ether of 4-acetylcyclohexene-1-carboxylic acid (44) seems to be the shortest one.

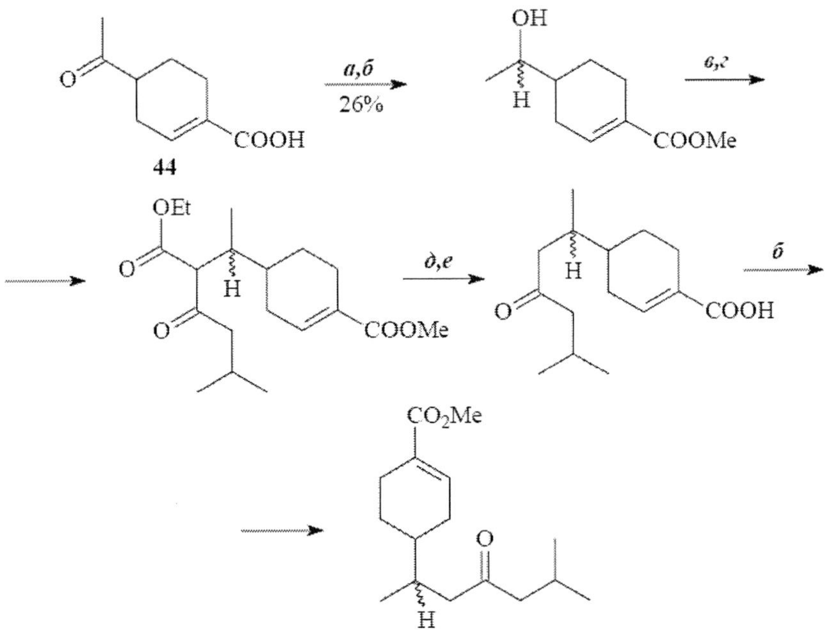

Reagents: *а*.$NaBH_4$; *б*.CH_2N_2; *в*.TsCl/Py; *г*. 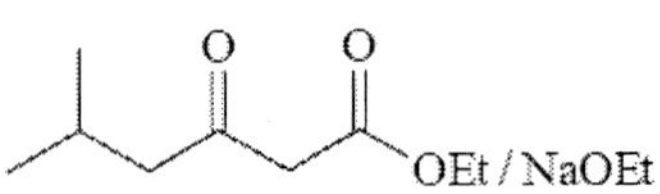
д. KOH/EtOH; *е*. Heating;

Stereospecific synthesis of individual (*R,R*)- and (*R,S*)-diastereoisomers of juvabione is made from (*R,R*)- and (*R,S*)-alcohols (45), which optical purity was increased by fractional crystallization of their 3,5- dinitrobenzoates.

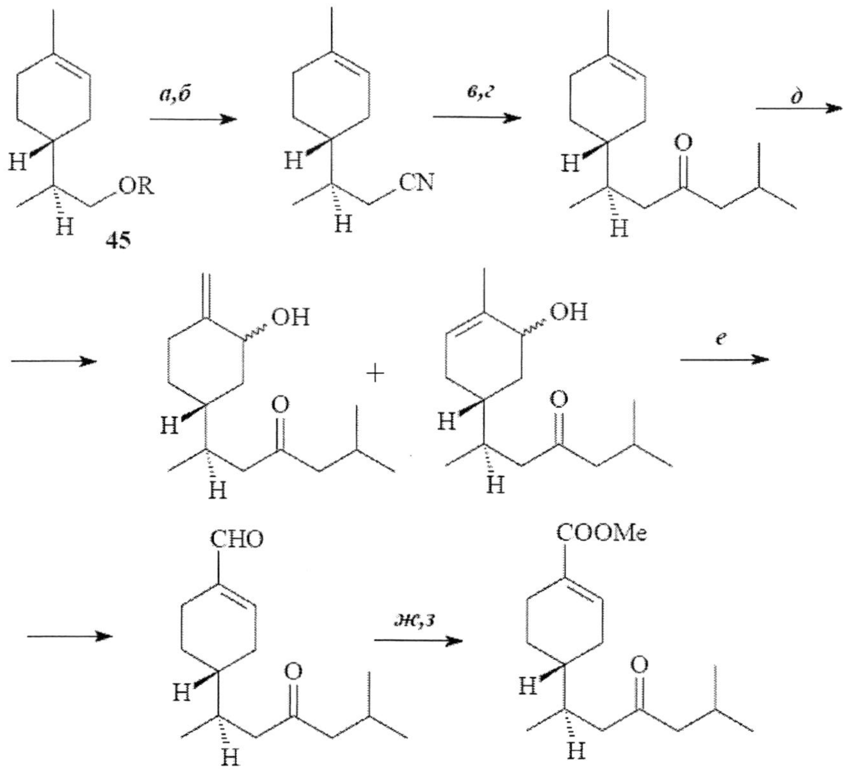

R=H, DNB Juvabione

Reagents: *а*. TsCl/Py; *б*. NaCN/DMS; *в*. Me_2CHCH_2Li; *г*. H_3O^+
д. O_2/hν; *е*. $Na_2Cr_2O_7/H_2SO_4$/HOAc; *ж*. $AgNO_3$; *з*. CH_2N_2

Pellitorine was synthesized in the following way.

- Piper nigrum is successfully applied as cardioprotective agent and as antitussive drug as a part of complex Ayurvedic preparation produced by Indian company Plethico Farmaceutical Ltd. «Travesil».
- Since ancient times American Indians have been using coneflower against infectious diseases. E. angustifolia is more often applied in medicine. In Europe E. purpurea is more studied. Coneflower is rich in polysaccharides and phytosterols making stimulating effect on immune system. Root E. Angustifolia contains glycoside echinacoside possessing antimicrobic action, oils of thyme and pyrrolisidine alkaloids (tussilagine and isotussilagine). Active components of herbs are caffeic and ferulic acids and their derivatives (chicory acid, and echinacoside). Polysaccharide echinacoside possesses antimicrobic, antifungal and cortisone-like activity contributing towards resistance (immune protection) against viral and bacterial agents and suppressing inflammations. Sesquiterpene ethers possess immunostimulating activity. Inulin as a part of coneflower increases activity of complement system. Roots and herb of a plant contain: betaine, rutin, phytosterol, flavonoid glycosides, saturated and unsaturated fatty acids. All parts of a plant

contain macro- and microelements: iron, zinc, selenium, potassium, calcium, molybdenum, silver, cobalt, nickel, barium, beryllium, vanadium, manganese. Fresh juice E. purpurea in vitro increases phagocytosis.

- Pharmacopeial drug containing extract E. purpurea is manufactured by Lek company, its name being Immunal. Polysaccharides included into Immunal stimulate marrowy hematosis, thereby increasing number of granulocytes in peripheral blood, and activating phagocytic systems of the organism. Thus, Immunal mechanisms of action include: phagocytosis activation; increment in functional activity of B-lymphocytes and immunoregulatory balance; induction of interferon synthesis. Immunal stimulates nonspecific mechanisms of anti-infective protection, due to which it is recommended for prophylaxis of influenza and acute respiratory viral infection to people with lowered resistance against infections (at psychoemotional stresses, intoxications of different origin); for accelerating wound healing; at diseases accompanied by secondary immune deficiency; chronic inflammatory diseases (prostatitis, arthritis, and others).

14.4. Synthetic Analogs of Juvenile Hormones

In 1961 farnesol and farnesal were found to possess high juvenoid activity compared with natural JH, and methyl ether and diethylamide of farnesil acid proved to be far more active.

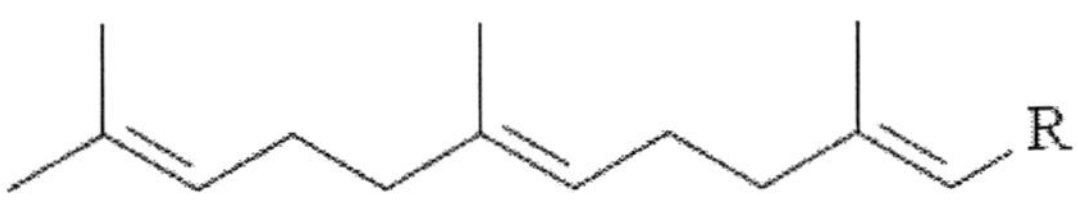

R= CH_2OH (farnesol), CHO (farnesal),
CO_2Me (methylfarnesoate),
$CONEt_2$ (farnesil acid diethylamide)

It resulted in vast researches aimed at searching for active JHA and their synthesis. As a result of large-scale searching several kinds of juvenoids were revealed – i.e. substances being analogs of natural JH. Highly active JHA are found, mainly, among acyclic compounds with isoprenoid fragments in chain to which derivatives of farnesil acids and geraniol ethers belong, in particular. The most active juvenoids are found in ethers of isoprenoid 2,4-diene acids (2,4-dienoate). Ethyl ether of 3,7,11-trimethyl-2,4-dodecadienoic acid, trade-named as hydroprene, isopropyl ether of 3,7,11-trimethyl-11-methoxy-2,4-dodecadienoic acid, known as methoprene and kinoprene being propargyl ether of 3,7,11- trimethyl -2,4-dodecadienoic acid proved to be the most perspective for practical application.

Many different methods for synthesizing these very active JHA are proposed. Hydroprene as mixture of (2*E*,4*E*) and (2*Z*,4*E*)-isomers was synthesized from partially hydrogenated citral (46) and ethyl bromoacetate by Reformatsky reaction.

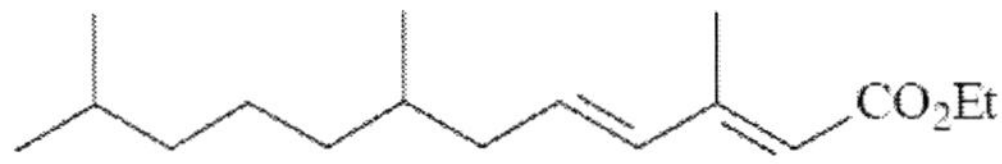

Hydroprene

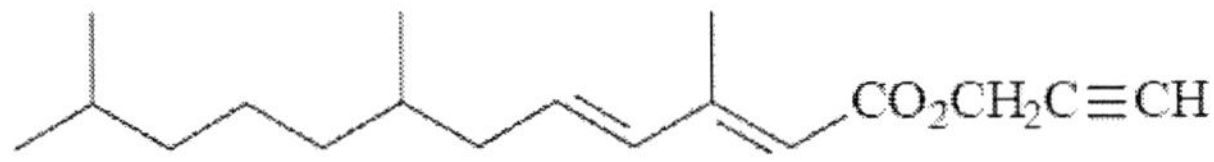

Kinoprene

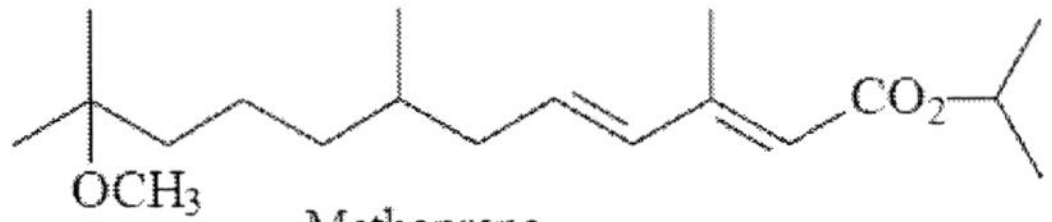

Methoprene

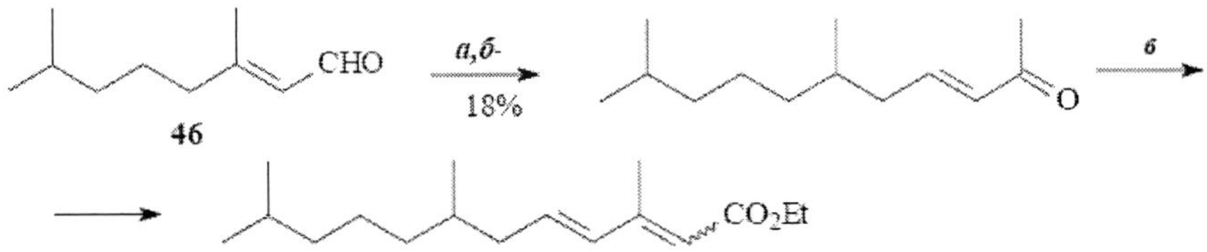

46

Reagents: *а*.Me_2CO; *б*.KOH/EtOH; *в*.$BrCH_2CO_2Et$/Zn;

- Subdivision of Degussa corporation specializing in production of fine chemistry products has finished its project for making new production line on plant of Jayhawk company amounting to USD 13 million. The enterprise absolutely meets chemical standards cGMP (current Good Manufacturing Practice) accepted in the USA. The declaration of successful technical inspection made by Food and Drug Administration (FDA) over a new enterprise and quality laboratory located in Galena (Kansas) was made by representatives of Degussa AG, with its head office being in Dusseldorf, Germany. The enterprise built for 3 years occupies the territory of 1000 sq. m and is planned for manufacturing biorational antiparasitic preparation Methoprene by order of Wellmark International. The management is made by Jayhawk Fine Chemicals (Galena, Kansas) which is a subsidiary of Degussa corporation. Flexible technological line of a new enterprise including multiple-unit reactor and dehydrator filter will make it possible to manufacture other similar hormonal insect growth regulators in future. The plant is also equipped with new control room, thermal exchange tower, water cooling system, transformer substation, air separating unit and water treatment works. As Methoprene manufacture is a complex multistage process the technological process from raw materials to finished products is fully automated using up-to-date distributed control system.

However this method is ineffective due to low yield of product of aldol-crotonic condensation and formation of great number of by-products.

The way of hydroprene synthesis proved to be successful based on using of available synthon received from petrochemical wastes – 4-methyltetrahydropyran (47).

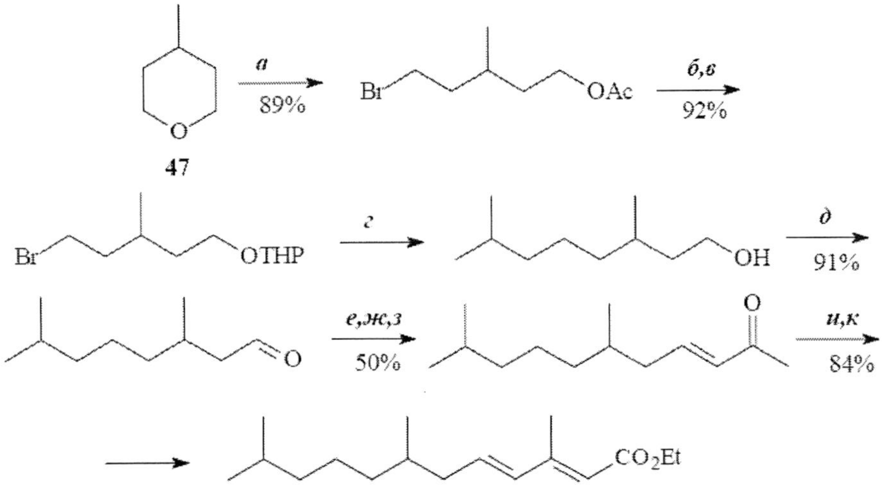

Reagents: а.AcBr/ZnCl$_2$ б.TsOH/MeOH; в.DHP/TsOH; г.Me_2CHCH_2MgBr
д.Py/CrO$_3$/HCl; е.CH_2=$CHCH_2MgBr$; ж.O_2/PdCl$_2$/CuCl; з.H_3O^+
и.BrMgC≡COEt; к.H_2SO_4

Action of some synthetic analogs of juvenile hormones was found not to be limited with morphogenetic effects only. Immediately after applying altosid (preparative form of isopropyl ether of (2E,4E)-3,7,11-trimethyl-11-methoxydodeca-2,4-dienoic acid (methoprene) – one of the most effective synthetic analogs – on cutaneous covering of silkworm larvae (*Bombyx mori*), contents of free amino acids within fat body is increased twice (Figure 58).

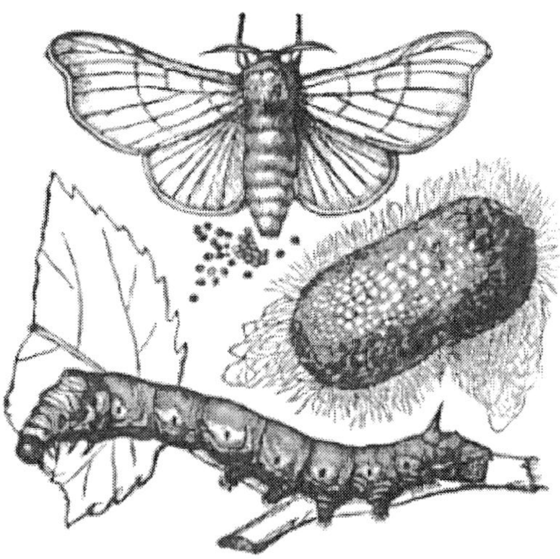

Figure 58. Bombyx mori.

Together with accumulation of additional protein matters within caterpillar body, the fifth larval instar preceeding formation of silk cocoon is prolonged. It results in significant increase of volume of raw silk output. Positive effect from altosid application in silkworm breeding would be higher if this compound is available. Full synthesis of hydroprene and methoprene is made under products of ozonolitic splitting of 1.5-dimethyl-1-cyclooctene (48) by scheme given below. Individual (2*E*.4*E*)-isomers are selected by preparative highly effective liquid chromatography.

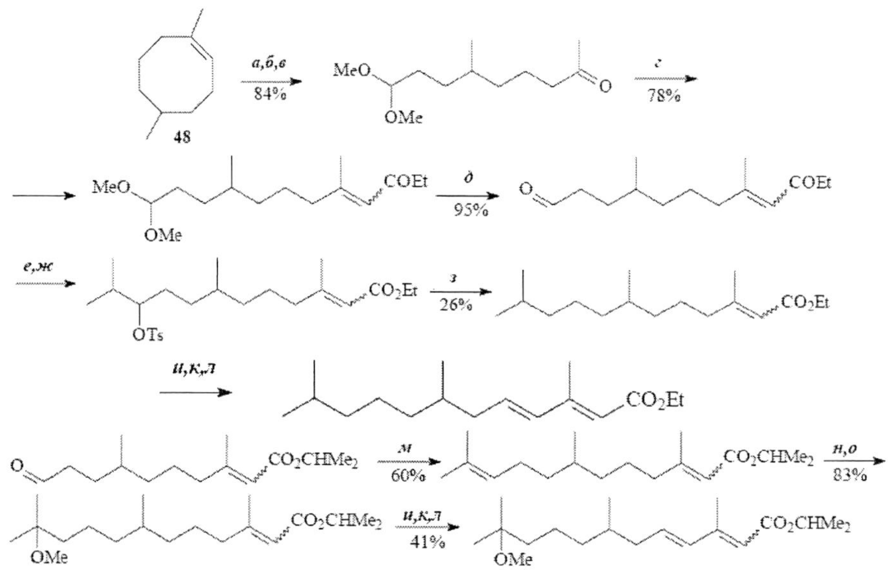

Reagents: *а*.O_3/MeOH; *б*.H_2/Pd-$CaCO_3$-PbO; *в*.NH_4Cl/MeOH; *г*.$Me_3SiCH_2CO_2R$/ iso-Pr_2NH; *д*.H_3O^+; *е*. $Me_2CHMgBr$; *ж*.TsCl/Py; *з*.Zn/NaI/THF; *и*.NBS/$(BzO)_2$; *к*.Li_2CO_3/LiBr/DMF; *л*.ВЭЖХ; *м*.Me_2C=PPh_3; *н*.$Hg(OAc)_2$; *о*.$NaBH_4$/NaOH

Simple and effective scheme for synthesizing 2.4-dienoates being analogs of natural juvenoid – pellitorine was elaborated. It made it possible to get wide range of juvenoids based upon available ketones:

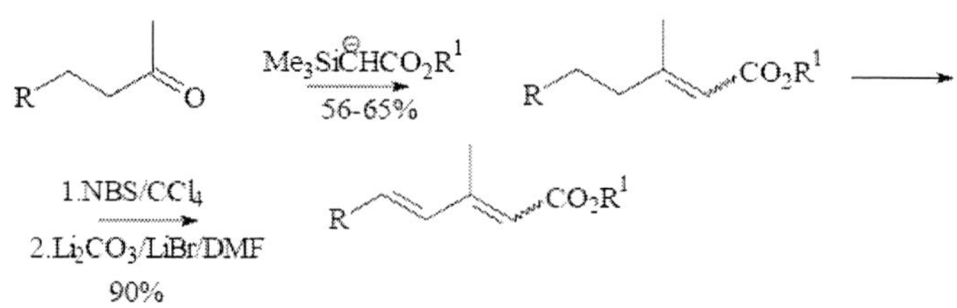

Modification of alcoholic part of a molecule was made by saponification of ethyl ether into 3-methyl-2.4-decadienoic acid with further treatment of its acid chloride (49) with appropriate alcohol. When using alkylamines and 1-aminopropionic acid the appropriate amides are received:

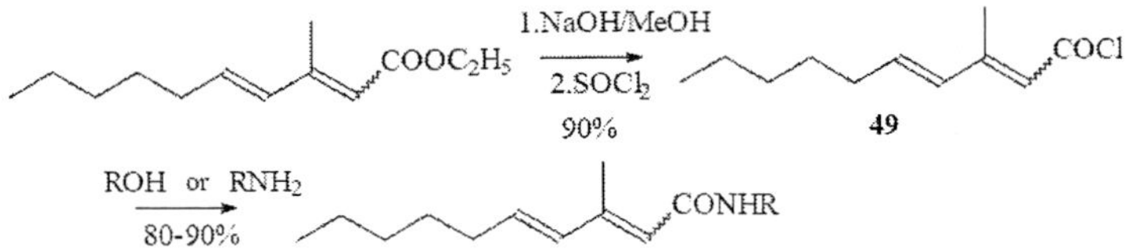

·Analysis of biological activity of ethyl ethers of 3-methyl-2,4-undecadienoic acids on other species of Coleoptera has revealed that this class of compounds is effective also against maize and granary weevils and saw-toothed grain beetle.

14.5. Ecdysone and Insects Hormones

Term "ecdysteroids" proceeds from Greek word "*ecdysis*" (molt) and unites in itself a group of polyhydroxylated sterols, with a structure similar to α-ecdysone i.e. molting hormone and metamorphosis in arthropoda [87].

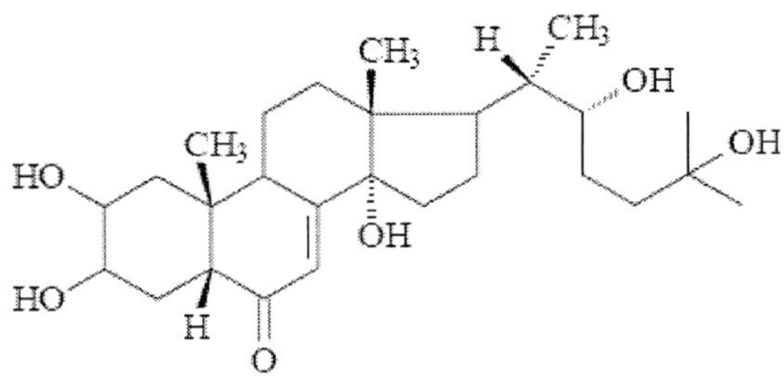

α-ecdysone

At present, phytoecdysteroids are found to be widely distributed secondary metabolites of plants. These substances structurally similar to ecdyson molecule (principal molting hormone of invertebrates) proved to be characterized by more variety of chemical structures than zooecdysteroids. By now, over 150 different structures of phytoecdysteroids have been identified. They form continuous series with other vegetable steroids similar in chemical structure: brassinosteroids, withanolides, steroid glycosides, and others. According to one of most substantiated hypotheses phytoecdysteroids are allelochemical* toxins and antifeedants for unspecialized species of phytophagan insects.

This point of view corresponds to general concept on role of substances of secondary metabolism of plants as protective factors. Phytoecdysteroids are seen as a part of multicomponent chemical protecting strategy appeared in the course of coevolution of plants and phytophagan invertebrates [88, 89, 90].

However, phytoecdysteroids are not universal protective factors of plants, they are effective against unspecialized species of insects only and are often found in plants together

* Allelochemical - (from Greek *allelon* one another, between one another, mutually) –chemical substances acting like data carriers between living organisms.

with other toxic allelochemical substances (e.g. ajugarines, withanolids, alkaloids, saponins, flavonoids), probably interacting with them in synergism* [91, 92].

Main metabolic predecessor of ecdysteroids in insects is cholesterol (C27-sterol). As insects cannot synthesize steroid nucleus *de novo**, they get sterols with food. Predatory species get cholesterol directly with animal food, and phytophages have to use vegetable sterols being homologs of cholesterol and differ from them by availability of additional methyl (campesterol) and ethyl (stigmasterol and sitosterol) derivatives at 24th carbon atom (accordingly C28- and C29-sterols). By enzymatic reactions of dealkylation the majority of arthropoda can transform C28- and C29- phytosterols into cholesterol. Content of ecdysteroids in some plants (e.g. in fern, brown pine *Podocarpus elatus* (Figure 59) and some Asiatic plants) is so high that they can be isolated from them in amounts enough for practical application.

Figure 59. Podocarpus elatus.

- Fruit fly *Drosophila melanogaster* (Figure 60) gets ergosterin necessary for ecdysone with food, which is provided to it from usual baker's yeast *Saccharomyces cerevisiae*, quickly appearing on sleepy fruit and vegetables.
- The only species of edible mushrooms not affected by worms and insect larvae are chanterelles (Figure 61). Their resistance is explained by absence of phytosterols in them in the form assimilated by insects [93].

---

* Synergism (from Greek synergy is cooperation, concord) – combined effect resulting in increase of aggrehate effect.

**De novo* – from the beginning.

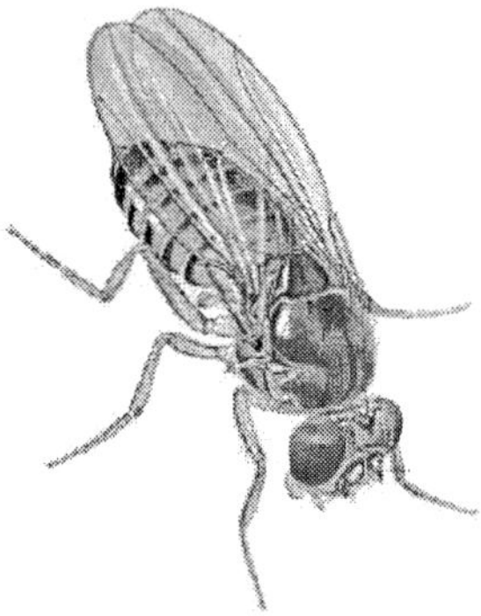

Figure 60. Fruit fly Drosophila melanogaster.

Figure 61. Chanterelle Cantharellus cibarius Fr.

Some species of insects (cutworm – *Heliothis spp*., Asiatic locust – *Locusta migratoria*) are resistant to very high concentrations of phytoecdysteroids in food (400-1000 ppm of 20-hydroxyecdysone), while others (cabbage white butterfly – *Pieris brassicae* and cabbage moth – *Mamestra brassicae*) reject food containing only 5-60 ppm of 20-hydroxyecdysone.

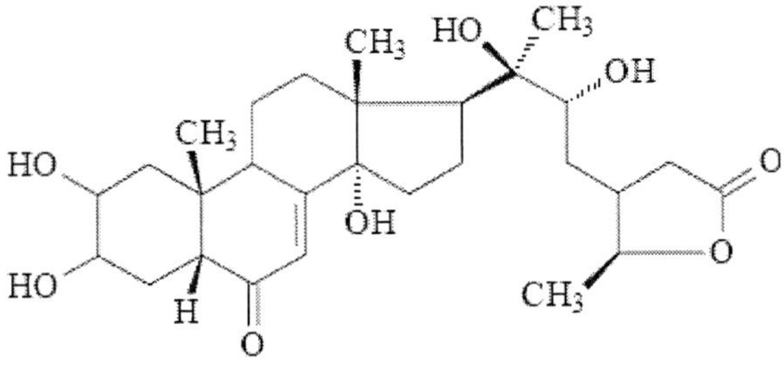

Cyasterone

- Cubo with his colleagues working in Kenya in 80-s noted that after devastating raids of locust flocks the only sort of bugle (*Ajuga remota*) among savannah vegetation survived. It has been shown that addition of this plant extract to food results in disturbance of normal insect development: when transforming from larva to pupa there appeared not one, but several head capsules. These extra capsules block mouthparts thereby resulting in inability to feed on and death from hunger.

- Ecdysteroids of crown sawwort lead to abnormalities in cabbage moth pupae. Cubo has defined that this development abnormality is caused by phytoecdysteroids contained in plants *A. remota*, which disturb transformation from larva into pupa. Each insect undergoes normal changes and cell movements preceding metamorphosis, however larval external skeleton was not shed.
- Bees as partners and allies of plants learned to use unique properties of plants, partially having lost their ability in adulthood to synthesize vital substances. They start to get them ready with nectar from wide range of herbs and woody plants. Thus, when there is no adequate and usual food, but only different substitutes or monofloral* honey, where necessary sufficient substances cannot be found, sharp weakening of insect organisms takes place. They become flaccid, weak and affected by different diseases. Similar state in bees is seen when affected by varroa mite, which sharply reduces ecdysteron level in bees necessary for vital functions and performance by exhausting hemolymph.
- In some animal species mating decreases lifetime, however underlying mechanism remained unknown for a long time. Dr. Jens Rolff and Dr. Michael Siva-Jothy managed to decipher it by the example of meal worms – beetles used as food for reptiles and birds. "More mating organisms, i.e. more successful from evolutionary point of view, thus reduce their life cycle", Dr. Michael Siva-Jothy explained. This discovery can be applied for agricultural pest control, which will be most vulnerable exactly after mating period.
- Canadian biochemists got X-ray images of specific proteins structure, due to which some insects can survive even at severe winters. These proteins bind the smallest ice grains at the start of crystallization thereby preventing their further growth. Similar substances save some cold-water fish, however their efficiency is 100 times less than protein properties of spruce budworms and meal worms. The scientists think the information received will be useful for making biological additions, improving quality and preservation of frozen foodstuffs.
- To get 25 mg of ecdyson, one had to make extraction from more than 500 kg of silkworms.
- There is striking similarity between stimulating action of prothoracotropic hormone on ecdyson secretion and action of adrenocorticotropic hormone on steroid formation by suprarenal glands.

14.6. Juvenoid Application

Methoprene manufactured commercially by a number of countries was widely applied in practice. It is especially active with diptera to which a lot of cattle ectoparasites belong [94].

* Monofloral – honey produced from one type of flower

Horn flies and face fly, cattle fly *Haematobia irritans* and biting housefly *Stomoxys calcitrans* lay their eggs to cattle excrements. Larvae grow and pupate also here, and blood-sucking adult species appear from pupae. They are able to transfer anthrax bacillus by biting. However, development of *H. irritans* and *S. calcitrans* in the manure can be suppressed by daily feeding cattle with encapsulated methoprene or directly with food. Eggs and horn fly larvae are especially sensitive to methoprene. Eggs and biting housefly larvae are less sensitive to methoprene [95, 96, 97].

Surface spraying of places of biting housefly *S. calcitrans* reproduction in pastures also gives good results in controlling it. Field tests have showed that far less methoprene is required for pest suppression as compared with chlorpyrifos, stirofos, ronnel and stirofos-dichlorvos mixture [98].

From their early development northern cattle grubs *Hypoderma bovis* and common cattle grubs *Hypoderma lineatum* bore in skin of cows and other ruminants, while growing and molting they move along animal innards, then again bore in and bite through skin, and within several months more live in the formed purulent hard carcinomas (Figure 62). Cows strongly affected with larvae lose their weight and reduce milk yield, hard carcinomas affect meat quality and depreciate skin dressing.

Double spraying of surface area of yearling calves naturally affected with warbles *Hypoderma spp.*, with 0.1% methoprene solution (in 36 days intervals) did not prevent from purulent hard carcinoma formation, but their size even expanded as compared with the original one. However, fallen out larvae of the last instar did not result in viable adult species after pupation. From biochemical and hematological points of view experimental calves did not differ from checked ones after methoprene treatment and no acute poisoning was seen [99].

Life cycle of sheep botfly *Oestrus ovis* can be disturbed, with its grubs being developed within sheep nasal cavities and frontal sinuses within 9-10 months causing serious inflammations (Figure 63, 64). Treatment of lamb nasal cavities with methoprene by 0.1-0.5 mg per lamb reduces larva pupation and suppresses adult species to 86-100%, although affecting no larva development [100].

As methoprene does not directly kill warbles and botfly larvae, its application in control of these pests is to be made under integrated system of their suppression.

In relation to egg-laying cattle mites (*Boophillus deoolonatus, B. nieropulos, Ambiliiomma nebracum*) methoprene possesses well-defined anti-gonadotropic* and ovicide* action [101, 102] that makes it perspective in control of these pests reducing milk yields and being able to transmit Texas fever. Methoprene can also suppress formation of adult species appearing from larvae of sheep pest – wingless fly *Melophagus ovinus* (sheep ked)[103].

The most perspective field of methoprene application is control of different pests damaging stored cereals, flour, flour goods, dried fruits, tobacco and other stored agricultural and food production.

In many countries there are pyralids which larvae damage cereals, flour, bran, rice, peas, whet grain, different seeds, dried vegetables and mushrooms, dried fruits, sometimes

---

* Gonadotropic hormones – animal and human hormones secreted by the pituitary gland and placenta that stimulate the gonads and control reproductive activity.

* Ovicide – chemical agent that kills eggs of pests and mites.

chocolate. Products affected by them are covered with cobweb, excrements and butterfly remnants. Other widely distributed pests are grain moths which caterpillars bore in grain and feed on until pupation, with grain weight reducing to 50%.

One-day pupae of Almond moth *Ephestia* (*Cadra, Anagasta*) *cautella* and flour moth *E. kuhniella* are more sensible to methoprene; hatching of adult species able for reproduction is suppressed by even small doses [104]. Dose of 25 mg/kg of methoprene fully suppresses appearance of adult species of Almond moth on seeds of unhulled peas, with residual methoprene being active for at least 8 months more after pea seeds treatment [105]. Indian meal moth *Plodia interpunctella* is more sensitive to methoprene: dose of 5 mg/kg of methoprene nutrient solution suppresses appearance of adult species of Almond moth by 93%, and of *P. interpunctella* - to 100%. Methoprene suppresses pupae development and affects diapause in southwestern corn borer *Diatraea grandiosella* [106, 107].

Figure 62. Hard carcinomas formed by ox warble fly larvae.

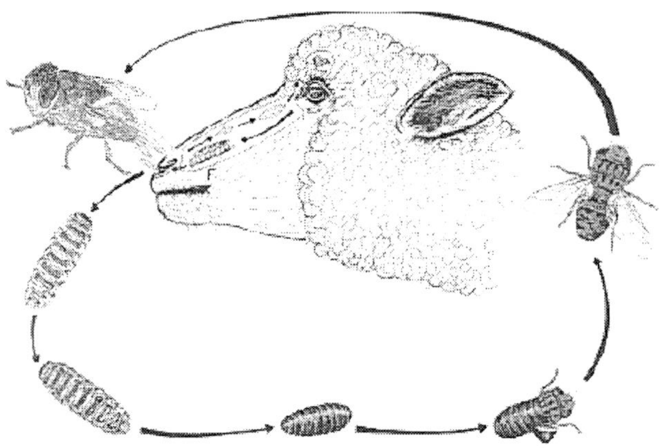

Figure 63. Botfly life cycle.

Corn moth way of life (larvae tunnel into grain) makes it difficult to control. Methoprene amounting from 1 to 10 mg per 1 kg of grain (applied for 5 weeks) is absolutely inefficient against Angoumois grain moth *Sitotroga cerealella*, presumably due to poor methoprene penetration into grain; hydroprene gives better results in controlling *S. cerealella* [108]. These two substances prolong period for postembryonic development of grain moth; organophosphorous insecticides like malathion are likely to be more effective in controlling it. Application of methoprene is more effective on fighting against rice moth *Corcyra cephalonica* – pest of foodstuffs in Southern Asia. Rice moth reproduction on sorghum grains is suppressed by methoprene in doses 5 and 10 mg/kg – accordingly by 82 and 100%; hydroprene proved to be more effective. However, when transmitted to unaffected grain majority of larvae pupate normally transforming into normal adult species with normal generation; this fact testifies to necessity of making repeated treatments [109].

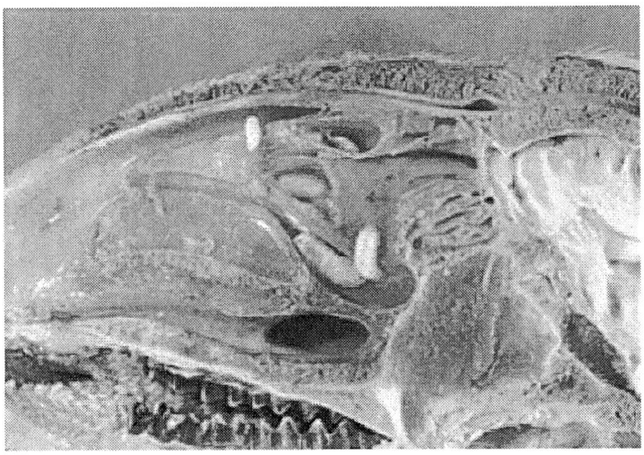

Figure 64. Botfly larvae.

As a rule, meal worms, dermestids, saw-toothed grain beetle, grain beetle, lesser grain borers, seed beetles and small cigarette beetles damage different stored foodstuffs both in larva and adult stages. Females of these species lay their eggs onto grain surface, dried fruits and on rough surfaces of packs, in the upper level of flour, etc. The damaged products become not only unfit or of little use, but in some cases even dangerous to health.

Other group of pest beetles – snout beetles, is the most dangerous pest of grain into which females lay their eggs. Hatched larvae grow, feed, pupate and transform into adult species inside the grain, young beetles also feed on grain contents (Figure 65). Grains of wheat, rye and corn after being left by beetles reduce their mass by 20-75%, lose germinating capacity and cause disorders, and inflammation of digestive system, if eaten.

To control first group of pests methoprene is more effective as compared with the second group where carbamat larvicide* R013-5223 is used [110]. Weevil population can be significantly reduces by making methoprene penetrate into grain (by increasing of dose, number of treatments, reducing aeration, etc.) [111].

* Larvicide – from English *larva* – a chemical used for killing larvae of insects and mites.

Confused flour beetles and especially red flour beetles are sensible to comparatively small doses. According to other data, methoprene concentration being 2-4 mg per kg of grain, decrease of number of adult species of confused flour beetles in wheat and corn reaches 100 and over 99% accordingly. Aftereffect of residual methoprene on *T. castaneum* reaches 12 months. Undesirable effect from treatment is appearance of large non-pupated larvae of additional instar (living up to 120-150 days), increase of life time of normal larvae of the last instar for several days more and hence, additional damage for the stored products.

Methoprene activity is small if applied to warehouse beetles *Trogoderma inclusum* and *T. variabile*; hydroprene in concentration of 5-50 mg per kg of medium gives better results: 100% suppression of breeding, 75-86% suppression of hatched larvae [112]. On the contrary, lesser grain borer *Rhizopertha dominica* is very sensitive to methoprene, and residual action of methoprene on *R. dominica* lasts up to one year [111].

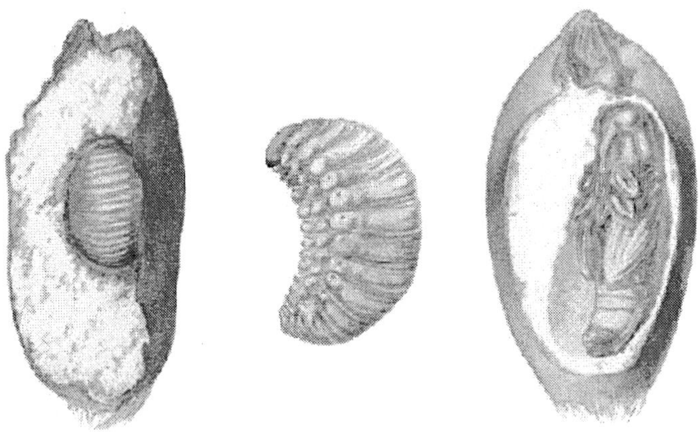

Figure 65. Granary weevil Sitophilus granarius feed on grain.

Methoprene is highly effective for suppressing cigarette beetle *Lasioderma serricorne*. In concentrations of 5 mg/kg of medium it not only fully suppresses breeding of adult species, but makes strong ovicide action. If exposed within 24 hours on a surface of a list treated with methoprene, female beetles fail to lay viable eggs, therewith breed population of treated females decreases pro rata power of methoprene concentration on the surface of the list [113]: average adult breed of one female decreases under control from 57.4 to 0.1 under methoprene concentration of 0.1 mg/cm^2. To control cigarette beetle in tobacco warehouses and in drying chambers in the USA they are sprayed with special 5% preparation form of methoprene - Kabat. Residual effect of methoprene after tobacco treatment lasts for up to 4 years. Another tobacco pest – warehouse (tobacco) moth *E. elutella* – is also suppressed together with cigarette beetle.

Methoprene doses of 1-5 mg/kg absolutely suppress reproduction of saw-toothed grain beetle *O. surinamensis* and related to it grain beetle *O. mercator* on wheat, oats, corn, bran, and doses of 5 mg/kg and more suppress larva hatching. Methoprene do little damage to larval stage of *Oryzaephilus* and its length actually remains the same. Granary weevil *S. granaries* is more sensitive to methoprene than maize weevil *S. oryzae*, however, methoprene application without reducing grain aeration does not provide economically approved level of suppression [114]. So, concentration of 50 mg/kg of medium suppresses breeding of granary weevil by 46%, and maize weevil only by 22%. However, when grain is stored in closed premises 100% suppression is achieved at even low concentrations, which is likely due to methoprene vapor penetration.

Methoprene also suppresses reproduction of cowpea weevil *C. maculatus* on alfalfa seeds: seeds soaking in aqueous emulsion with concentration of 100 mg/l and higher prevented from hatching of adult species on treated seeds within 4 weeks [115].

Application of methoprene against pests of stored products has advantages over fumigation (which effect does not last long) and over treatment of stored products with usual organophosphorous and carbamate insecticides (which are more toxic). In closed warehouses photodecomposition of methoprene under the sunlight is practically excluded, and rate of microbiological degradation at low temperatures of products storing is reduced. Residual effect of methoprene can last for several months and even years [116]. As a rule, rate of methoprene application is ten and hundred times lower than of usual pesticides.

Great interest for practical usage are 3-methyl-2,4-dienoates. Analysis of biological activity of ethyl ethers of 3-methyl-2,4-decadienoic and undecadienoic acids on other Coleoptera representatives have revealed that this class of compounds is also effective in control of maize, granary weevils and saw-toothed grain beetles. Insect breed was reduced by 41, 61 and 34% accordingly in 3 months after treatment, while methopren (reference preparation) reduced breed of maize and granary weevils by 19% only. Activity against maize weevil is even higher (60%) for ethyl ether of 3-methyl-2,4- decadienoic acid [117].

Chapter 15

15. Biosynthesis Inhibitors of Insects Chitin

Important class of new generation insecticides is 1-phenyl-3-benzoylurea, acting as biosynthesis inhibitors* of chitin (BSIC) of arthropoda. Their lethal action is in their oral or topical (through outer coverings) penetration into insect organism, thereby preventing formation of chitin protecting delicate tissue against environmental factors [118]. BSIC are selectively toxic to arthropoda and safe for warm-blooded animals and humans (LD_{50} = 2500-8500 mg/kg) [119, 120]. By rates of application they are similar to pyrethroids.

First information on insect BSIC of chlorine derivatives activity of 1-benzoyl-3-phenylurea refers to compounds (50, 51) [121-124]. Further intensive research revealed great number of benzoylphenylurea with far more expressed inhibitory action, e.g. compound (52).

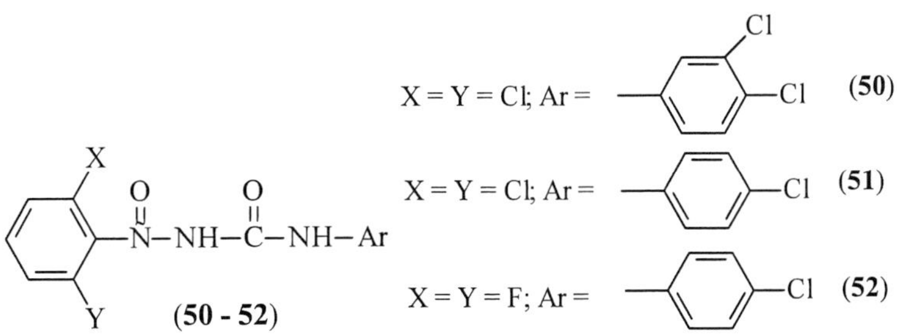

Widely applied 1-(2,6-difluorobenzoyl)-3-(4-chlorophenyl)urea, or diflubenzurone (52) is manufactured abroad under the trade name of "Dimilin" [125]. Dimilin penetrates through chitin of many insect species suppressing embryonic development in egg inside an insect additionally revealing ovicide action. These properties of dimilin accompanied by low toxicity made it main preparation in controlling many agricultural insects.

Analysis of biological activity of analogs of 2,6-diflubenzurone (52) allowed some regularities to be discovered. Character of substitutes in *ortho*-positions of aromatic ring is found to significantly influence inhibitory activity (IA). Growth of activity in row 2,6-dihaloaroyl derivatives 50-52: [X = Cl, Y = H] < [X = Y = Cl] < [X = Y = F] was revealed

* Inhibitor (Latin *inhibere* – to retard, to stop), substance that inhibits the process, reaction.

among great number of urea derivatives of the considered class. When electron-accepting hydrophobic substitutes are introduced into aroyl fragment of a molecule, inhibitory activity of urea is increasing, and with increase of substitute volume, inhibitory activity is decreasing [126]. Degree of dihedral angle formed by aromatic ring plane and amid group of benzoyl fragment of a molecule was found to significantly influence inhibitory activity. Evidently, due to formation of intramolecular hydrogen bonding of carbonyl of amid group with *ortho*-hydrogen atom providing complanarity of benzoyl fragment, inhibitory activity of derivatives with one *ortho*-atom of fluorine is essentially lower than for 2,6-difluorobenzoyl derivatives, where angle between planes of aryl and CONH is optimum (~30° as per [127]). Substitution of benzoyl group to heterocycle (pyridine or piperidine) does not result in increase of inhibitory activity [128]. Electron-accepting lipophilic substitutes in the 4th position of aromatic ring promote to increase of insecticide activity of a compound. Such substitutes include fluorine, alkoxycarbonyl, alkoxy and perfluoroalkyl groups. Additional introduction of chlorine atoms into 3rd and 3.5-positions also increases inhibitory activity of a compound.

On the contrary, substitution of phenyl fragment to pyridine one increases insecticide activity of urea [129-133]. Cycles with heteroatoms [134], distinct from nitrogen, reduce inhibitory activity [135, 136]. Inhibitory activity reduction is also noted in derivatives of thiourea [137, 138]. Introduction of substitutes over nitrogen atoms negatively affects inhibitory activity of urea derivatives [139, 140, 141]. Very high inhibitory activity is achieved under introduction of phenyl ring of alkoxypyridine grouping with acceptor substitutes (compound 53) into *para*-position. Substitution of carbonyl group into thiocarbonyl - compound (54) or sulfonyl one - compound (55) is found to result in decrease of effect from biosynthesis inhibition of insect chitin. Availability of bond aroyl-urea-aryl is mandatory condition for high insecticide activity. E.g., sharp decrease of inhibitory properties is noted for compounds (56-60). Modification of urea fragment due to introduction of C=N-bond (compounds 61, 62) increases acute and chronic toxicity of BSIC [142].

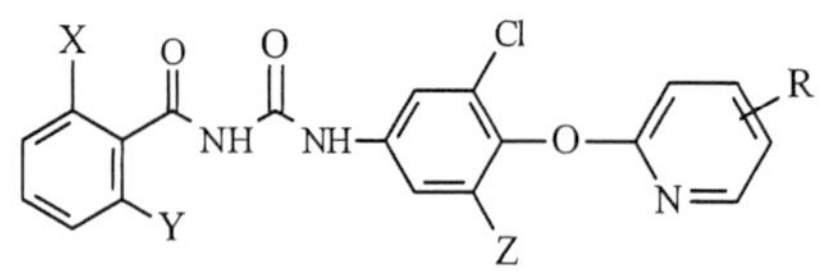

53

X = H, Cl, F; Y = Cl, F; Z = H, Cl; R = NO_2, CN, CO_2Alk

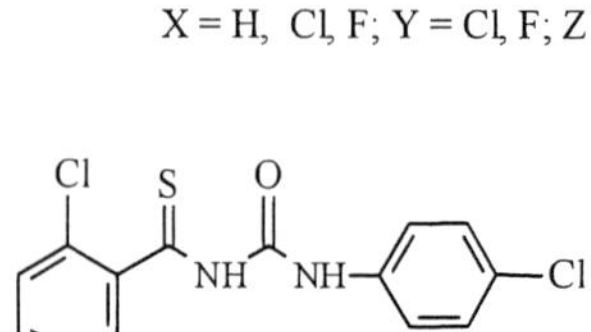

54

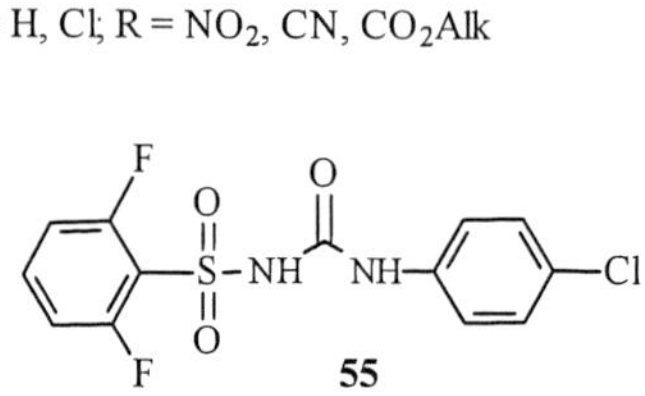

55

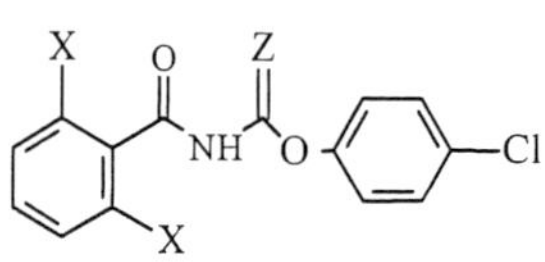

X = F; Z = O (**56**)
X = Cl; Z = O (**57**)
X = F; Z = S (**58**)

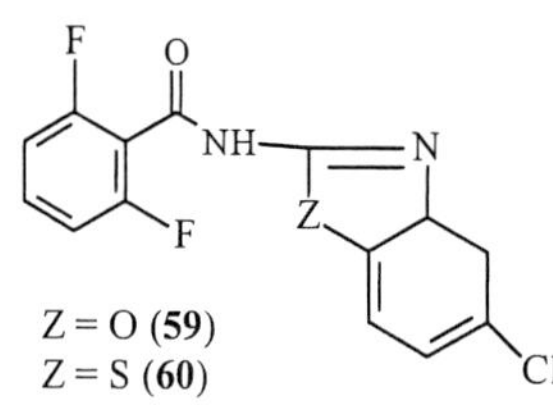

Z = O (**59**)
Z = S (**60**)

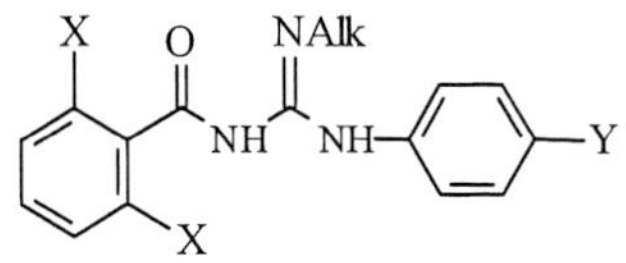

61

X = Y = Cl
X = F, Y = Cl
X = Y = F

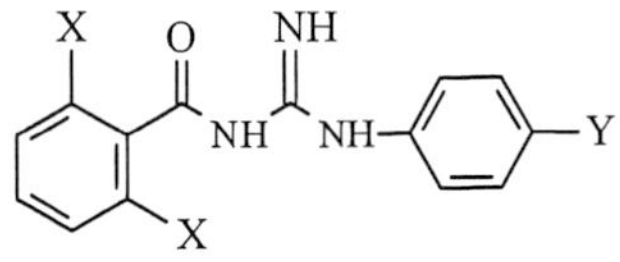

62

X = Y = Cl
X = F, Y = Cl
X = Y = F

Range of BSIC is widely expanded due to sulfur-containing analogs of diflubenzurone [143, 144, 145] (63-70).

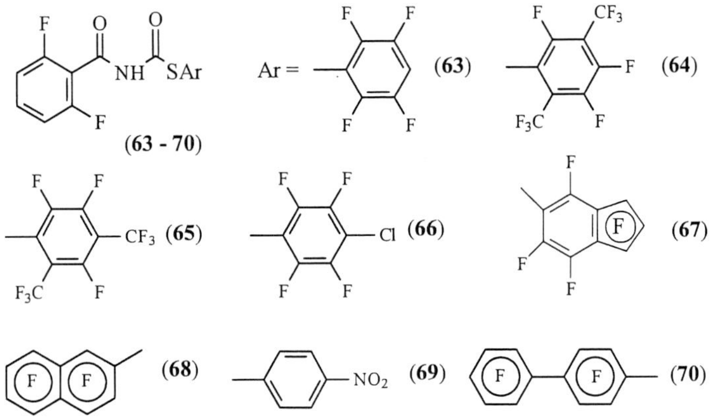

(63 - 70) (63) (64) (65) (66) (67) (68) (69) (70)

New analog of 2,6-diflubenzurone (71) revealing high BSIC-activity is found [146].

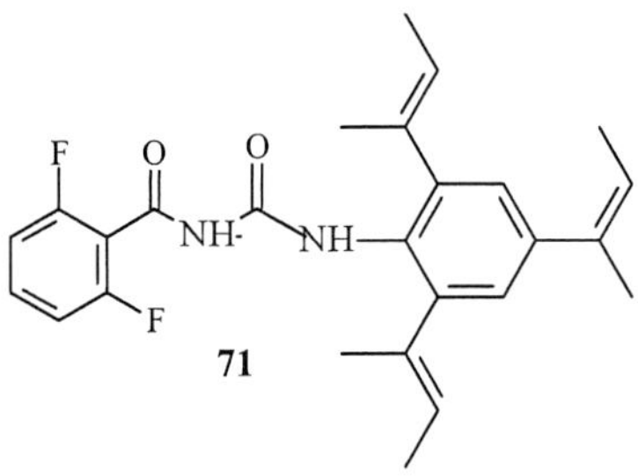

71

15.1. Synthesis of 1-Phenyl-3-Benzoylurea

Interaction between benzoylisocyanates (73) and arylamines (74), or, appropriate benzoylamids (75) with arylisocyanate (76) is more often used for synthesis of phenylbenzoylurea (72) [147, 148, 149]

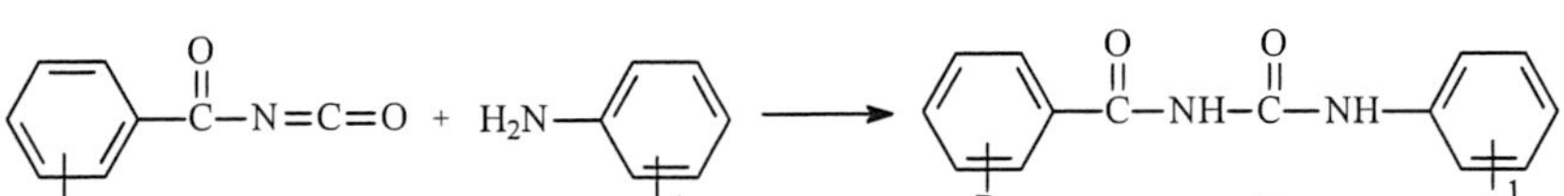

73 74 72

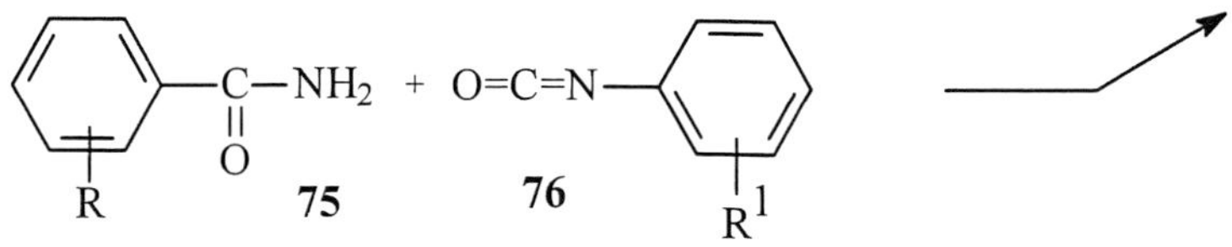

75 76

R = R^1 = H, Hal, Alk

Alternative approach to synthesis of acylurea which is interaction between 2,6-difluorobenzamide (77) and crabamoylchloride (78) or ethers of arylcarbamino acid (79) is less known [150, 151].

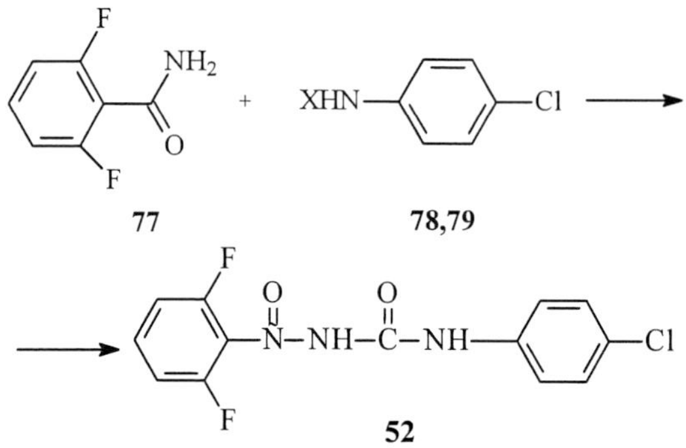

77 78,79 52

X = COCl (**78**), CO_2R (**79**) (R = Me, Et)

2,6-diflubenzurone (52) is received with 95% yield at carbonylation of mixture of difluorobenzamide (77) with *para*-chloronitrobenzene (80) [152].

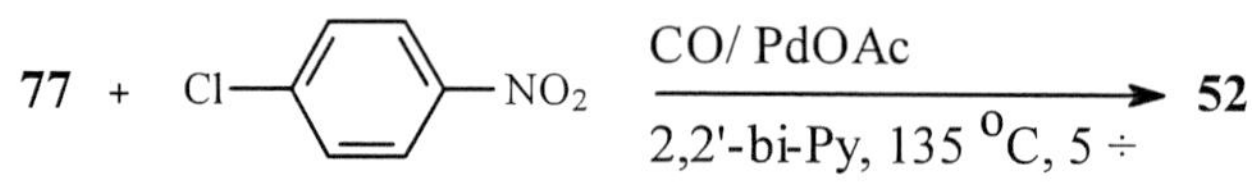

80

Effective synthesis of acylurea (83) using arylamids (82) and trichloroacetamides (81) recovered easily from appropriate amines and trichloroacetic acid influenced by phosphorus trichloride is proposed [153, 154].

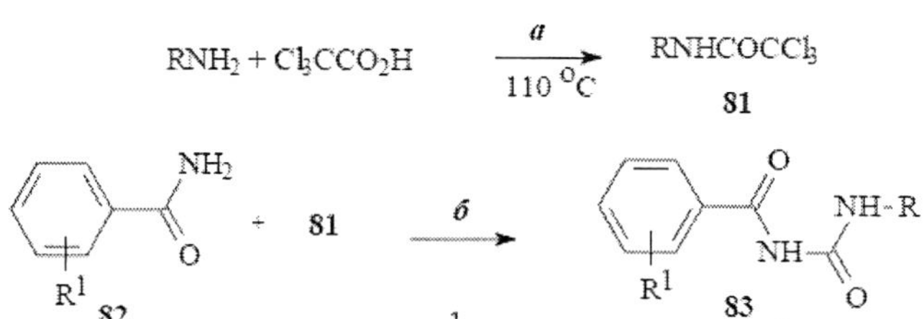

81 82 83

R = Alk, Ar; R^1 = Alk, Hal

Reagents: *a*. PCl_3 *б*. NaOH/DMSO

Analysis of known approaches to synthesis of 1-phenyl-3-benzoylure demonstrates that method based on benzoylisocyanate is the most universal.

2,6-Difluorobenzoylisocyanate (84) is usually obtained by interaction between 2,6-difluorobenzamide (77) and oxalyl chloride [155, 156]. Other convenient way for synthesis of 2,6-difluorobenzoylisocyanate (84), including interaction between 2,6-difluorobenzoylchloride (85) and trimethyl silyl isocyanate over $SnCl_4$ is found.

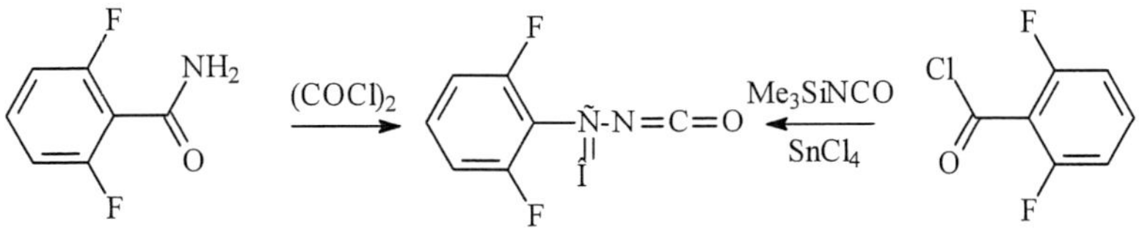

77 **84** **85**

By using this method 2-(trifluoromethyl)-6-fluorobenzoylisocyanate (87) – synthon for new analogs of 2,6-diflubenzurone [157] was synthesized from 2-(trifluoromethyl)-6-fluorobenzoylchloride (86) [158]

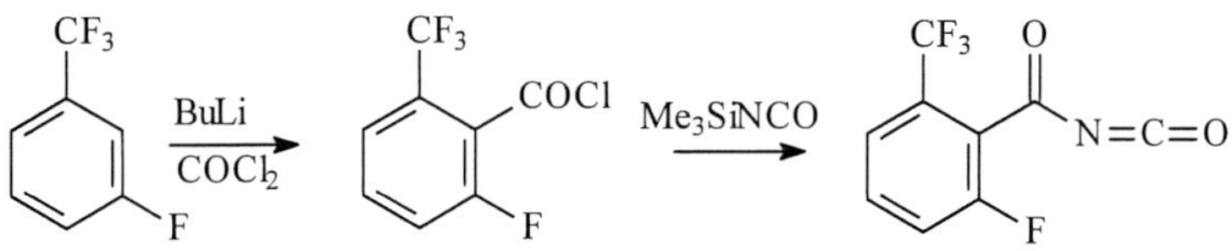

86 **87**

2,6-Diflubenzurone (52) and some of its analogs can also be synthesized by interaction between 2,6-difluorobenzamide (77) and arylisocyanate. Efficiency of method for getting 2,6-difluorobenzamide (77) is important for this approach. E.g. it is efficient to get it by consistent treatment of difluorobenzene (88) with sodium amide and methyl chloride. 2,6-Difluorotoluene (90) obtained in such a way is converted into benzamide (77) through intermediate 2,6-difluorobenzotrichlorides (90) and 2,6-difluorobenzoyl chlorides (91)[159].

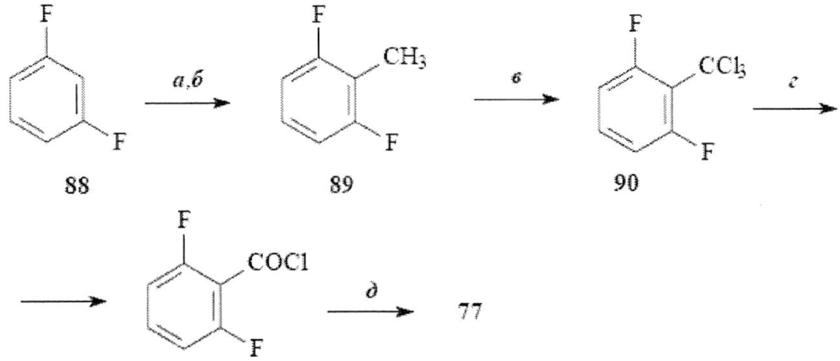

88 89 90

91

Reagents:а.$NaNH_2$; б.CH_3Cl; в. Cl_2/ hν; г.ZnO; д.NH_4OH

Aggregate yield of amide (**77**) is significantly increased, if methallyl chloride is to be applied instead of methyl chloride on alkylation stage. The resultant 2,6-difluoro-(2-methyl-1-propene-1-il)-benzene (93) is isomerized in refluxing toluene over TsOH in 2,6-difluoro-β,β-dimethylstyrene (94). Alkene ozonolysis (94) and further treatment of peroxide product of ozonolysis with hydroxylamine hydrochloride result in 2,6-difluorobenzaldoxime (95), regrouped under Beckmann reaction into benzamide (77) [160].

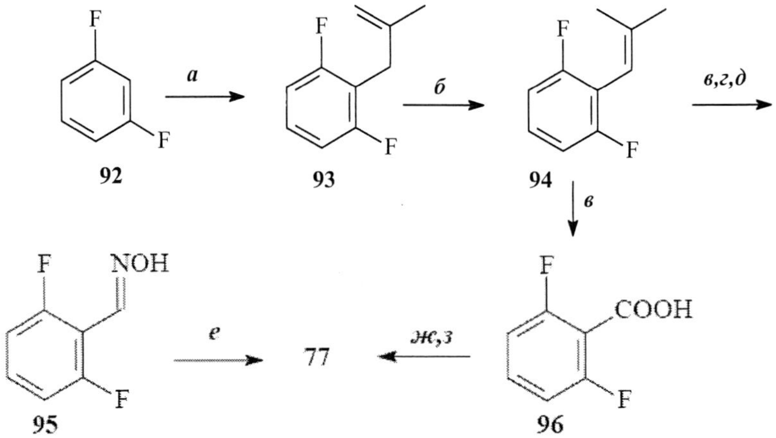

Reagents: *а*. 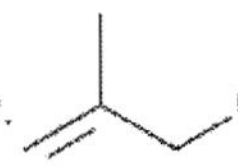 $Cl/NaNH_2$; *б*.TsOH; *в*.O_3/MeOH; *г*.$NH_2OH{\cdot}HCl$;

д.Na_2CO_3; *е*.H_2SO_4; *ж*.$SOCl_2$; *з*.NH_4OH

Benzamide (77) can be also obtained from 2,6-difluoro benzoic acid (96) synthesized in its turn by oxidation of 2,6-difluorotolyene (89) with ozone over cobaltous acetate (III).

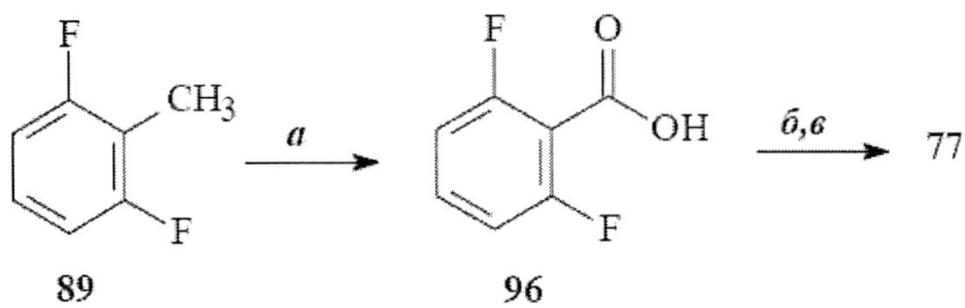

Reagents: *а*. $O_3/Co(OAc)_3$; *б*.$SOCl_2$; *в*.NH_4OH

Chapter 16

16. Pheromones and Insect Repellents

16.1. Insect Pheromones

Pheromones* are substances, secreted by insects for intraspecific communication: searching for mate, swarm, scattering, etc. As a rule, insect pheromones are volatile enough organic substances secreted into the environment as sprays affecting other insects in far distances from the source of smell. Knowing pheromone composition, which can include several individual chemical substances (major and minor components), it is possible to deliberately confuse insects, thereby affecting their normal development, mating, food searching, etc. However, wide application of these low toxic agents for protecting plants, materials, raw materials, environmental safe for the warm-blooded organisms, is limited by complexities in their extraction from natural sources. Today the only source of their extraction in quantities enough for practical application is chemical synthesis.

Pheromones are most promising first of all for control of pests of cereals and other foodstuffs, and due to safety application and effectiveness. At present pheromone structure of great number of grain and grain product pests is defined, and lots of different schemes of synthesis thereof are elaborated.

Sex pheromone of flour moths *P. interpunctella* and *Anagasta kuehniella* [161] - 9Z,12E-tetradecadiene-1-ylacetate (100) has been synthesized by using (*E*)-stereoselective Julia reaction and (Z)-stereoselective Wittig reaction proceeding from methylcyclopropylcarbinol (97) through intermediary homoallyl bromide 98 and ether 99.

- Flour moth larvae *Anagasta kuhniella* secrete into environment (flour, cereals, bran) the whole spectrum of 2-acyl-3,6-dihydrooxy-2-cyclohexen-1-ones containing oleic and other unsaturated fatty acids as acyl group. These substances affect larvae population density in the substrate causing their crawling away (like antiaggregation pheromones) and at the same time attracting insect pest *Venturia canescens* (Figure 66) (like Kairomones).
- Among cactuses there are examples of allelochemicals acting like Kairomones. Drosophila *Drosophila pachea* is attracted by schottenol generated by cactus

* Pheromone – from Greek *pherein* – to transmit and *horman* – to excite, stimulate.

Lophocereus schotii; this sterol acts like attractant. On the other hand, lophocereine and pilocereine alkaloids contained in cactus act like repellents against other species of drosophila thereby providing *D. pachea* with wide uncompetitive life area.

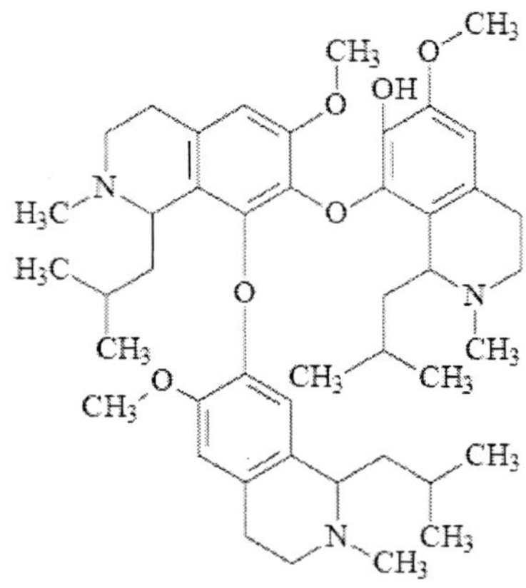

Pilocereine

Insect pest
Venturia canescens

Figure 66. Venturia canescens.

Selective ozonolysis of trisubstituted double bond over thermal disubstitute in olefin 101 made it possible to get tosylate 102, which oxidation by Yokker-Tsudzi resulted in ketotosylate 103, selective transformations of the latter gave component of aggregation pheromone of lesser grain borer 104 *Rhizopetrha dominica* [162].

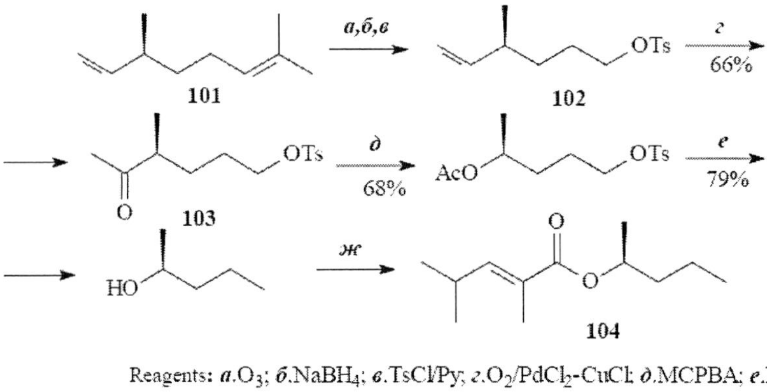

Reagents: *а*.O_3; *б*.$NaBH_4$; *в*.TsCl/Py; *г*.$O_2/PdCl_2$-CuCl; *д*.MCPBA; *е*.$LiAlH_4$;

ж.

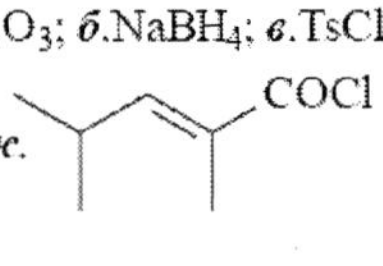

Chiral alcohol ester obtained by simple transformations from (R)- pulegone is used in synthesis of (4S,8R)- and (4S,8S)-dimethyl decanal 105 revealing attractive activity to confused flour (*T. confusum*) and red flour (*T. castaneum*) (Figure 66) beetles.

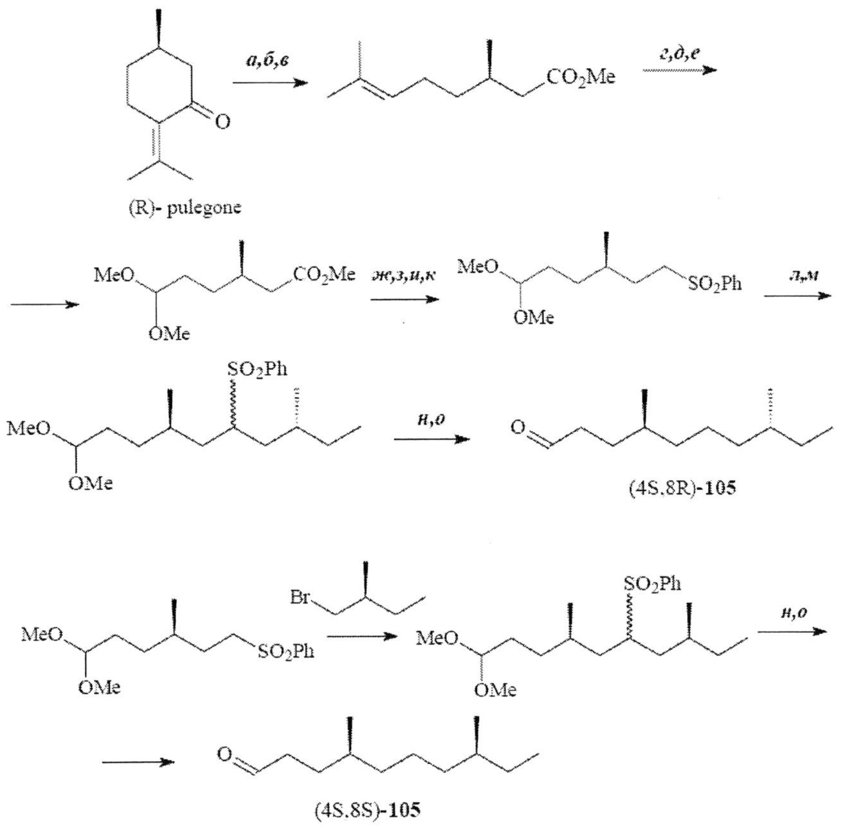

Reagents: а.HCl; б.NaOH; в.TsOH/MeOH; г.O_3/MeOH; д.Me_2S; е.TsOH/MeOH; ж.$LiAlH_4$; з.TsCl/Py; и.NaI/$NaHCO_3$; к.$PhSO_2Na$; л.n-BuLi; м.  н.Na-Hg; о.H_3O^+

Based on citronellyl tosylate (106) both stereoisomers of 4-methylnonan-1-ole are synthesized of which (*R*)-enantiomer (107) is a sex pheromone of yellow mealworm beetle (*T. molitor)*

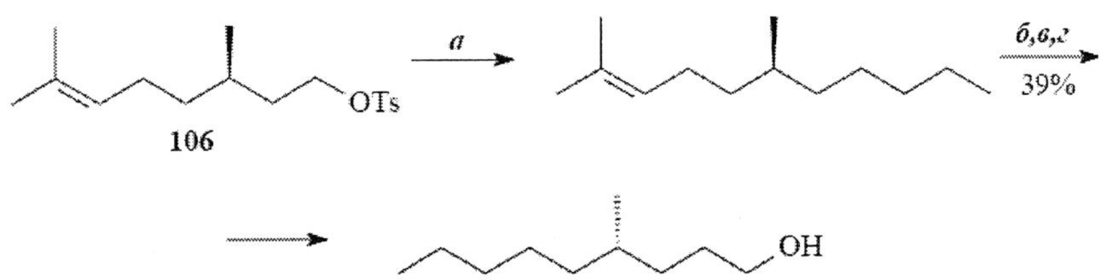

106

107

Reagents: а.n-PrMgBr/Li_2CuCl_4; б.4-dimethylaminopiridine; в.HIO_4; г.$LiAlH_4$

Sex pheromone of Dermestidae *T. variabile* contains 4-hexanolide (112). Pheromone 112 is synthesized from synthon 111 obtained by two-stage transformation of cyclic unsaturated ketone 113. Aldehyde-ether 110 is received from cyclopent-2-en-1-one (108) by ozonolitic splitting into methyl ether of 4,4-dimethoxybutanoic acid (109) with further hydrolysis of acetal group to aldehyde influenced by pyridinium tosylate. Condensation of aldehyde-ether 110 with Grignard reagent with high yield gave secondary alcohol 111.

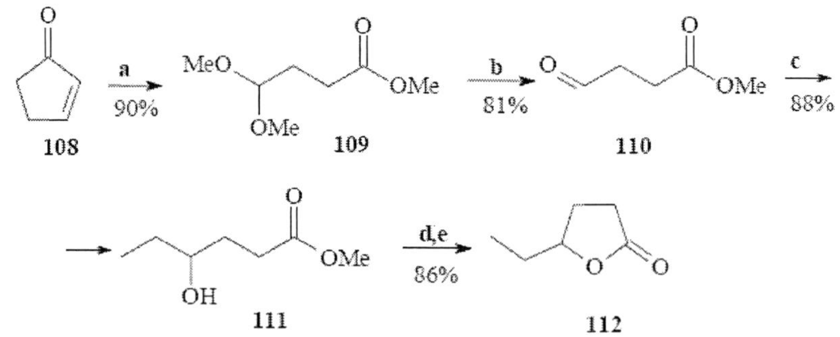

Reagents:: a.O_3/MeOH/p-TsOH; b.Tos Py/Me_2CO; c.EtMgBr/Et_2O; d.KOH/MeOH e.HCl

Further, methyl ether of 4-oxyhexanoic acid (111) is easily and entirely transformed into specific pheromone 112 influenced by KOH in MeOH, to be further treated with 1% HCl. Yield of pheromone 112 at cyclization stage made 86%.

In synthesis of racemic forms of *cis*-trogodermol, (±)-116, and *trans*- trogodermol, (±)-117, corresponding to two minor components of pheromone compositions of dermestids (Figure 57) *Trogoderma*, (±)-1-bromine-5- methylheptane (114) was used as the "left" part of a molecule. Carbon skeleton was composed by alkylation of terminal triple bond, and *Z*- or *E*-geometry of final products was defined by restoration of triple bond in alcohol 115 which is general for predecessor (±)-116 and (±)-117.

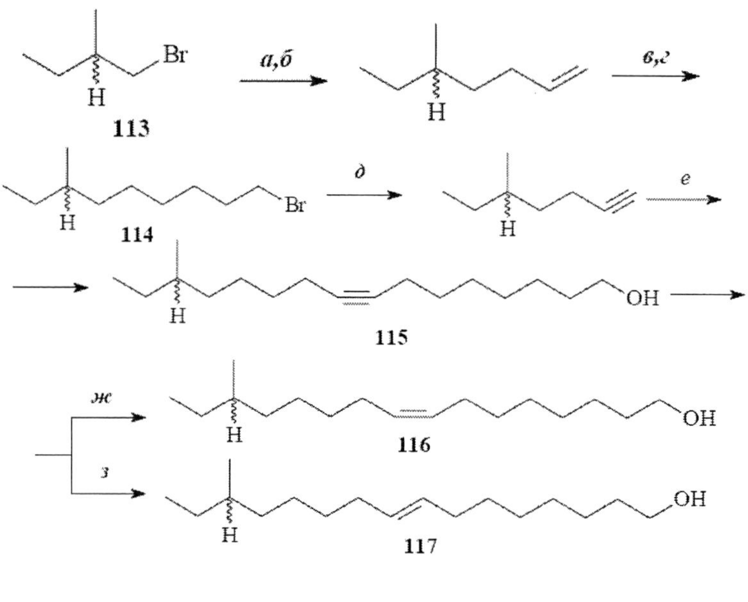

Advantage of the first one is simplicity of operation, and that of the second one is technologic way for obtaining bromide 114 from available original compounds.

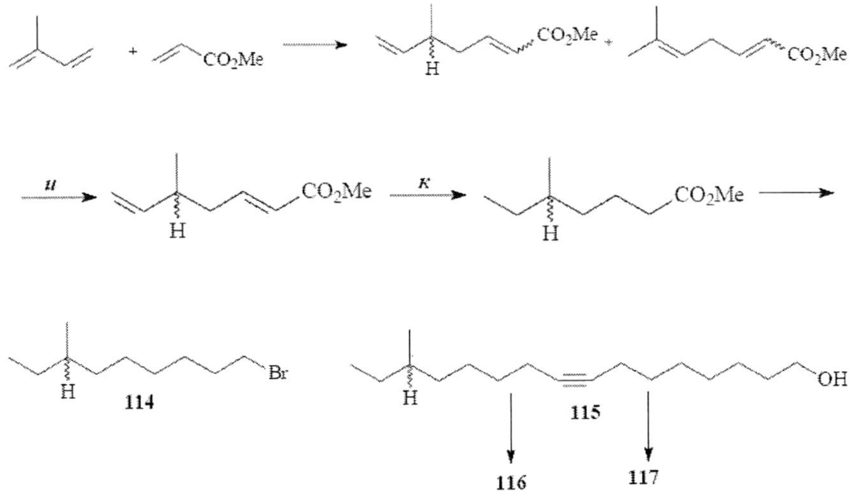

Reagents: *а*.Mg/Et_2O; *б*.CH_2=$CHCH_2Br$; *в*.B_2H_6-THF, then H_2O_2/NaOH; *г*.PPh_3; *д*.NaCCH/HMPA; *е*.LiNH/Cl$(CH_2)_6$OH; *ж*.H_2-Ni_2B(P-2)/EDA; *з*.$LiAlH_4$-$(MeOCH_2CH_2)_2O$, 140 °C; *и*.rectification; *к*.H_2-Ni/MeOH, 70 °C, 100 atm; *л*.$LiAlH_4$/Et_2O; *м*.LiCC$(CH_2)_7$OLi/THF-HMPA; *н*.H_2-Pd/$BaSO_4$-Kinone;

Pheromones belonging to a group of chiral compounds are known for bloodsucking biting housefly *Stomoxys calcitrans* and for two species of tsetse fly (Figure 67 – kaodzera carriers.

Figure 67. Tsetse fly.

- Midges are collective name for blood-sucking diptera, including gadflies, mosquitoes, midges and biting midges. Midges are small insects of gray or black color with short legs and short proboscis.
- Veterinary importance of midges is defined by their role as mass, harmful bloodsuckers and carriers of agents of some animal and human diseases. Attacking

animals they crawl inside eyes and ears, penetrate through upper airways. Bites at first insensitive later result in itch, burning, painfullness, oedemata, local and general temperature rise. Besides, midges consume much blood. Midge saliva entering animal organism is quite poisonous, as it contains hemolytic toxin. Simuliotoxicosis is progressing in animals at mass attack of midges. Animals experience much anxiety therewith, and then depression. Salivation* and oedemata appear, pulse and breath become rapid, lymph nodes are expanded. Animals move little, lie much, tetanillae appear. Intoxication develops fast, death occurs within 1-2 days. All animal species are attacked by midges, but cattle and horses are mostly affected. As a rule, simuliotoxicosis is a mass disease, with mortality among the diseased reaching 10-20%. The diseased animals significantly reduce in weight, milk yield in cows being reduced up to 50%. Besides, blood-sucking midges are specific intermediate hosts for onchocercosis pathogens in animals and humans. They can mechanically transfer pathogens of such extremely dangerous diseases like anthrax, encephalitis and tuleremia.

Universal and practical way for synthesizing pheromone of biting houseflies *S. calcitrans* is based on partial ozonolysis of isoprene cyclooligomers.

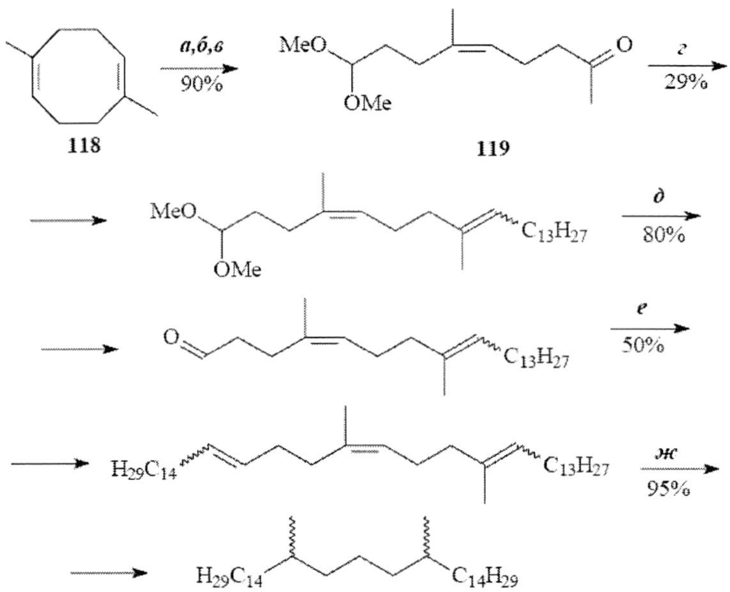

120

Reagents: *а*.O_3; *б*.H_2-Pd; *в*.MeOH/NH_4Cl; *г*.n-$C_{14}H_{29}P^+Br^-$/BuLi, -30 °C; *д*.H_2O-Me_2CO/PPTS; *е*.n-$C_6H_{13}P^+Ph_3Br^-$/BuLi -60 °C; *ж*.H_2-Pd/C-EtOH

* Salivation – the secretion of saliva by salivary glands. Big gland salivation takes place reflectorily by irritating sensitive nerve endings of mouth cavity with food, or when affected with conditioned excitators (appearance, odor of food).

So, ozonolysis of 1,5-dimethylcycloocta-1,5-diene (118), reduction of ozonides and acetalization lead to ketoacetal 119. Further deprotection of aldehyde group, repeated olefination by Wittig and exhaustive hydrogenization of trienoic hydrocarbon obtained therewith results in housefly pheromone 120.

3*Z*,6*Z*,8*E*-dodecatrien-1-al (122) – pheromone of termite *Reticulitermes virginicus* is received by olefination of 2*E*-hexenak (121) with phosphorane, generated from 3-butyne-1-bromide, with further oxyethylation of terminal acetylene with ethylene oxide (CH_2CH_2O) and by hydrogenation of the formed β-hydroxyacetylene.

Reagents: *а*. $HC{\equiv}CCH_2CH{=}PPh_3$; *б*. $LiNH_2/CH_2CH_2O$; *в*. $H_2/Pd\text{-}CaCO_3$

Olefination of aldehyde 123 with allyltitanic derivative generated from 1-(*tret*-butylthio)-3-(trimethylsilyl)-1-propane leads to diene 124 interacted enantioselectively with Grignard reagent, catalyzed with nickel(II)-1,3-bis(diphenylphosphine)propanedio-chloride. After alcohol oxidation (125) and olefination of the resulted aldehyde, 2E,6Z,8E-trienoic ether (126) is obtained by isomerization of Δ^2-bond and further hydride reduction transformed into 3*Z*,6*Z*,8*E*-dodecatrien-1-al (127) – trace pheromone of subterranean termites.

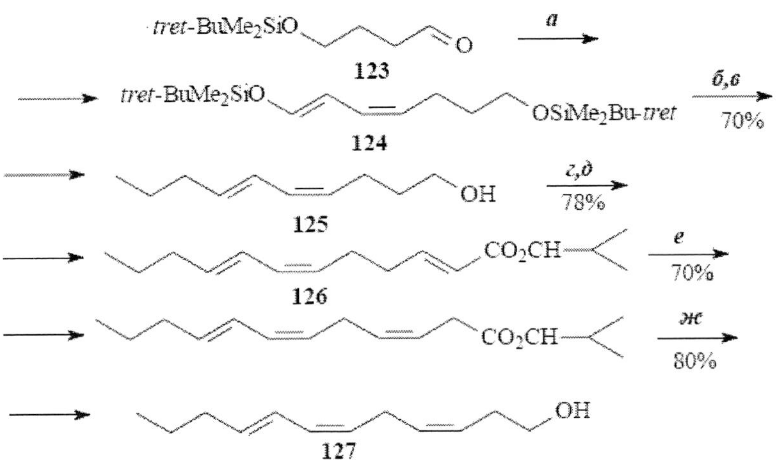

Reagents:: *а*. tret-BuSiCH=CHCH$_2$TMS/tret-BuLi/Ti(Oiso-Pr)$_4$;
б. PrMgI/Ni(dppp)Cl$_2$; *в*. Bu$_4$NF/THF; *г*. NCS-Me$_2$S;
д. (EtO)$_2$P(O)CH$_2$CO$_2$CH-isoPr$_2$; *е*. (TMS)$_2$NK; *ж*. LiAlH$_4$

11*Z*-octadecenal (129) – pheromone of a bee-moth *Achroia grisella* severely damaging apiculture was synthesized by using Wittig reaction based upon methyl ether of 10-undecenoic acid (128).

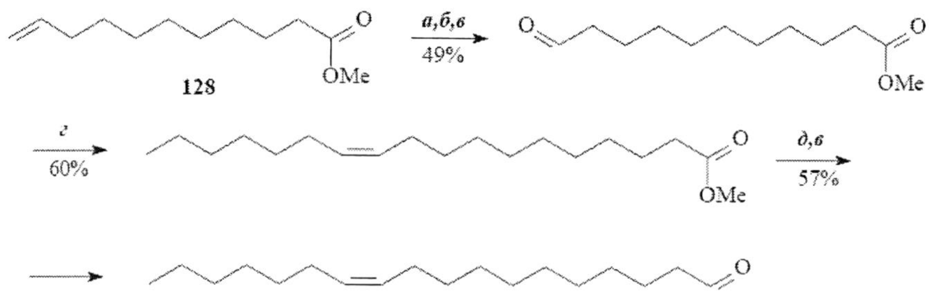

129

Reagents: а.B_2H_6/THF; б.H_2O_2/NaOH; в.PCC/AcONa/CH_2Cl_2; г.$Me(CH_2)_5CH{=}PPh_3$/THF; д.$LiAlH_4$/Et_2O;

- Larvae of greater and lesser bee-moths (wax moth) are of great interest for pharmaceutical industry as raw materials. Extract of wax moth larvae represented their infusion on a special component extracting useful substances has been used for more than 100 years. Application of this extract with unique properties forwards great number of different reduction processes in human organism. For example, numerous cases were noted in restoration of reproductive performance in women who could not become pregnant for a long time, as well as in stimulation of sperm motility in men. This product is also useful in treating catarrhal diseases, in problems with biliary excretion system, at cardiovascular diseases, and for normalization of arterial tension. In pre-Revolutionary period the extract was applied for tuberculosis treating - Koch's bacilli in human body form wax-like protective capsule, and enzymes of wax moth larva dissolve the wax.
- Males and females of a wax moth *G. mellonella* exchange with ultrasonic signals-clicks. On hearing sounds of the female, the male secretes pheromone for exact direction.

Stored sunflower seeds are strongly damaged by caterpillars of two species – Indian meal moth *P. interpunctella* and warehouse meal moth *E. elutella* After leaving the egg caterpillars feed on one seed then crawl to another one eating away embryos and therewith joining achenes into lumps with their sticky secretions. To control moths in sunflower seeds the following measures are recommended: humid disinfection of seed in transport lines and fumigation thereof with methallyl chloride or methyl bromide, making no negative influence on germinability. Humid disinfection is made with contact action pesticide: carbopfos, phoxim (volaton), ambush, actellic (5-15 g/t). For protective treatment of storehouses humid disinfection is made with decis, zolone, carbopfos, phoxim, actellic (0.2-0.4 g/m^2). Treatment of sacks with seeds of sunflower with such repellents as DBP (dibutyl phthalate) or DNP (dinonyl phthalate) is considered as effective protective measure. Repellent application protects seeds against moth butterflies. This method of pest control is regarded as useful and environmentally safe. To control moth butterflies it is recommended to treat sacks with seeds of sunflower with 0.5% (by preparation) process solution of decis of 2.5% emulsion concentration. It was found that 10 days after sack treatment all the butterflies completely died, and moth butterflies died in 30 minutes. Within 84 days after treatment the butterflies seen on the sacks died in 30 minutes, and on 114th day they died completely in one hour. In control (over sacks untreated with decis) natural mortality of butterflies took place on $7\text{-}9^{th}$ day. Due to its long-lasting toxic action (over 3 months) decis is considered the highly efficient preparation against Indian meal moths and warehouse moths [163].

Khapra beetle occupied wide area due to its polyphagy and high adaptation to new habitat factors. In new conditions the khapra beetle loses its primary food and adapts to new one. The analysis has revealed that khapra beetle larvae actively respond to different food substrates. Unrefined sunflower oil proved to be the most attractive food among tested one for the beetle larvae. The majority of the khapra beetle larvae was found in all sets of tests in meshes with sunflower oil, on average, about 25% of total number of larvae placed into central meshes. Crashed corn is also highly attractive for the khapra beetle larvae (20% of larvae chose this substrate); barley malt corn (16% of larvae), peanut and sunflower seeds (15 and 14% accordingly), hazelnuts proved to be less attractive - only 10% of larvae crawled into meshes with hazelnuts. Another peculiarity of the khapra beetle is its step-by-step habituation to fumigants. The research has revealed that the khapra beetle larvae survived at two-hour fumigation with methyl bromide by LD_{50} acquired resistance to it, which in the 6th generation was 1.81, and in the 16th – 2.11 [164].

Many substances stimulating oviposition by insects also simultaneously stimulate feeding on, or serve as attractants for insects. For example, sapogenin contained in soya beans delays eggs laying by Chinese weevils *C. chinensis* [165].

- Few species of bacteria belong to luminescent ones. Chemiluminescent reaction directly accompanied by luminescence is catalyzed by the enzyme — bacterial luciferase and includes processes of oxidizing reduced flavine mononucleotide FMN-H2 to FMN and simultaneously – of aliphatic (C14) aldehyde to myristic (C14) acid. This reaction is likely to take place through formation of flavine mononucleotide peroxide.

$FMN\text{-}H_2 + E + O_2 \rightarrow E\text{-}FMN\text{-}H_2\text{-}OOH$ $RCHO\text{-}E\text{-}FMN\text{-}H_2\text{-}OOH \rightarrow RCOOH + E\text{-}FMN\text{-}HOH \rightarrow RCOOH + E + FMN + H_2O$ + photon (490 nm)

Here, E - luciferase, OOH - hydroperoxide group, RCHO - aliphatic aldehyde, RCOOH – fatty acid formed under aldehyde oxidation.

Biochemical analyses are becoming more popular for the last years which apply whole bacterial cells (in suspension), extracts of luminescent bacteria, isolated enzyme – luciferase as test-objects. Adding $FMN\text{-}H_2$ to isolated bacterial luciferase, the researcher gets highly sensitive system to define aliphatic aldehydes. Insect sex hormones, pheromones, belong to their number, in particular. These pheromones are found in amount of 10^{-14} moles, which makes it possible to analyze metabolism of these substances in one species.

16.2. Repellents

Repellents* are natural and synthetic substances, repelling insects, mites, rodents and birds. Repellents are more often applied against bloodsucking insects (mosquitoes, gadflies, midges, etc.).

2-ethylhexane-1,3-diol (LD_{50} for rabbits 2.6 ml/kg) is used as repellent against bloodsucking insects. More often it is used together with other compounds.

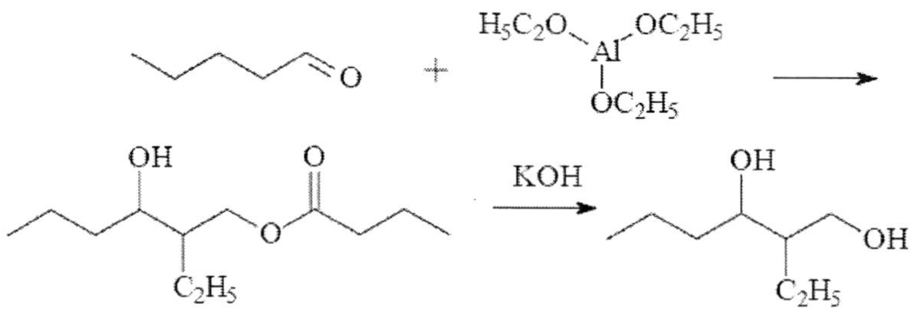

2-ethylhexane-1,3-diol

Acetoxime is proposed as preparation for clothes mothproofing [166]. Another oxime - cyclopentanone is patented as reactant of mothproofing preparation [167].

Tabatrex (dibutyl succinate) is used to repel bloodsucking flies from cattle, as well as to repel cockroaches and ants. Dibutyl adipate is repellent for bloodsucking ticks – carriers of encephalitis pathogen. It is applied for treating clothes of people working in taiga [168].

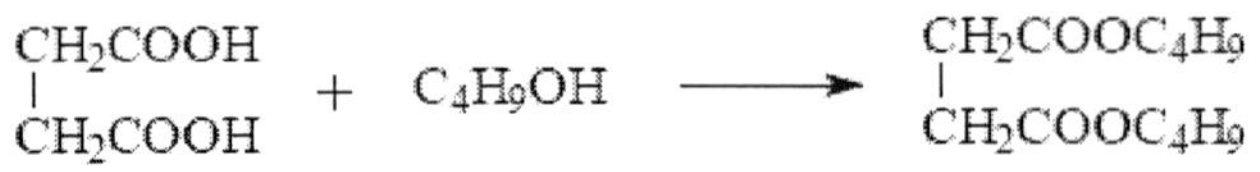

dibutyl succinate

Oxaminic acid ethers (ethane diacid monoamide) (130) and etheroamides of lower carboxylic acids display insect-repellent action. Among these compounds increase of hydrocarbon chain length up to C_8 results in rise of activity. Preparations with R= C_7-C_8 are most effective.

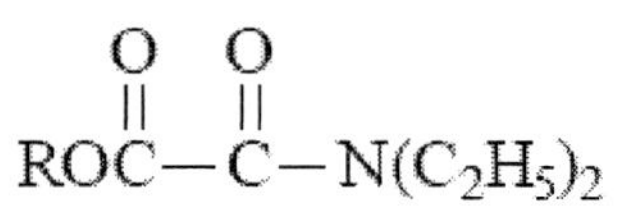

130

Oxamate is a mixture of alkyl ethers of N,N-diethyl oxaminic acid. The preparation is low-toxic, LD50 for mice is over 15 g/kg, it is easily destroyed and is not accumulated in animal waste products.

* Repellent (from Latin *repellens*) — distasteful or repulsive.

To control bloodsucking insects in cattle breeding amitraz is applied, with bisamidine being its effective agent.

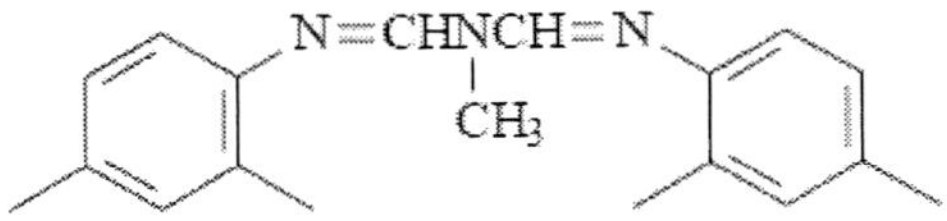

Bisamidine

Active repellents are also found among amides of chlorobenzole acid. N,N-Diethyl-m-toluamide (DETA) is widely applied as repellent of bloodsucking insects. It is water insoluble colorless liquid with LD_{50} of 2000 mg/kg. It is obtained under diethylamine treatment with methyl-benzole acid over catalysts.

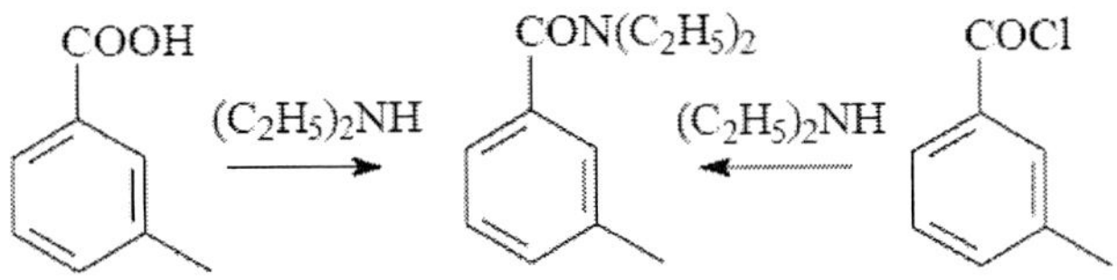

N,N-Diethyl-m-toluamide

Methyl-benzole acid is obtained by oxidation of *meta*-xylene with air over catalyst.

N,N-Diethylamide of 2,5-xylic acid known as DECSA possesses strong repellent activity.

Dimethylcarbate– dimethyl ether of *cis*-bicuyclo [2.2.1] hepte-2-en-5,6-dicarboxylic acid is applied as one of components in mixture with other substances to repel insects.

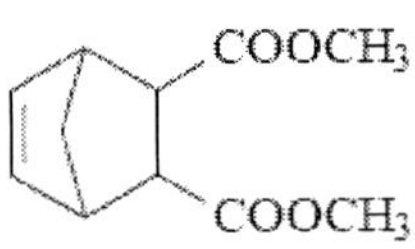

Dimethylcarbate

Pyran derivative [6-butoxycarbonil-3, 4-dihydro-2, 2-dimethylpyranone-4 (butopyron-oxyl)] is applied as repellent of bloodsucking insects in composition of complex preparations, which includes dimethyl phthalate and 2-ethylhexanediol-1,3 together with butopyronoxyl. Butopyronoxyl is insoluble in water, it is mixed with majority of organic solvents; no allergic reactions if applied to skin even exposed for a long time, LD_{50} is 7400 mg/kg.

Butopyronoxyl

Tee tree oil *Melaleuca alternifolia* which has been used as natural insecticide for a long time is applied as repellent for protection of a horse and a horseman against bloodsucking insects. Tee tree oil also possesses wound healing and anti-inflammatory action. There are a number of trees of *Leptospermum*, *Melaleuca*, *Kunzea*, *Baeckea* genus and others in Australia. These trees got generic name "tee tree", as sailors of captain Cook (who discovered Australia) used leaves of some bushes of genus *Leptospermum* from this family with lemon fragrance for tea infusion. Under steam distillation of leaves with yield of about 0.18% essential oil of yellow color and nutmeg flavor is received. Essential oil from this sort of manuka is quite often applied in medical practice. Essential oil received from different forms of *Melaleuca alternifolia Cheel* contains 1,8-cineole in different amounts. Main type contains up to 10% of 1,8-cineole, type «A» - 31-45% and type «B» 54-64%. Average commercial contents of 1,8-cineole is 5-10%.

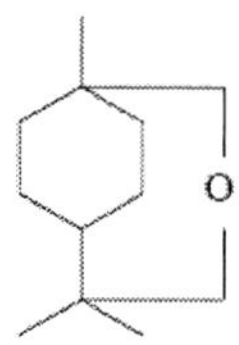

Cineole (1,8-epoxy-*n*-menthane)

Essential oils of local flora plants such as lavender, mint and basil are widely applied to repel and disorientate moth butterflies.

- Substance isolated from wild tomatoes is able to repel many insect species, including mosquitoes, mites and bloodsucking midge. Researches made by Dr. Michael Roe from the North Carolina State underlay a patent to use natural repellent in commercial products. The substance has already been tested for toxicity and is used in makeup preparations. In opinion of the authors of researches the substance is as effective as widely applied synthetic repellent DEET. The University has licensed rights to produce the substance of Insect Biotechnology, which financed the researches together with the National Science Foundation (NSF) and Agricultural Research Center of North Carolina.

Farmers from India and some regions of Eastern Africa traditionally apply dry leaves of basil (*Ocimum*) (Figure 68) to protect stored food from insects. Extracts *Ocimum kenyense, Ocimum suare, Ocimum kilimandscharicum* contain 1,8-cineole,. Extracts of these plants applied to covers of beetles *Sitophilus granarius, Sitophilus zeamais* lead to insect death. Eugenol and camphor also displayed strong repellent action.

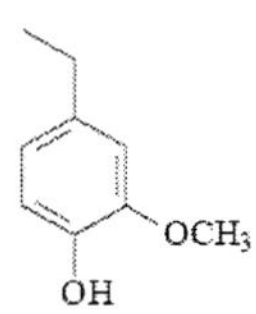

Eugenol

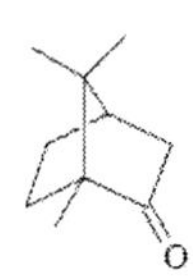

Camphor

Figure 68. Basil Ocimum basilicum.

Fruits and leaves of neem tree (*Azadirachta indica*, family *Meliaceae*) (Figure 69) have been used as repellent and natural insecticide for many centuries. Limonoids and a group of tetranortriterpenoid are active components of neem tree. Azadirachtin is one of the studied components of this plant, it is active to many species of insect-pests. When neem tree oil components contact or get inside the insect, its growth and development stop. By its action mechanisms it is like ecdysone and it disturbs complex hormonal balance.

Figure 69. Neem tree twig.

Azadirachtin is not toxic for pollinating insects as the preparation basically gets inside the insect through food. When azadirachtin is applied in high concentrations it has growth-regulating effect on young bees in small beehives, however in large bee-families such effect on adult bees is not noted. Residual insecticide activity remains within 7-10 days after plant treatment.

Azadirachtin acute toxicity on rats is LD_{50} 3.540 mg/kg. Two insecticides as "Azatin" by *AgriDyne* company and "Margosan-O" of *W.R. Grace and Co* have been elaborated in the USA. Both these products are azadirachtin extracts and contain some other limonoids intensifying preparation activity.

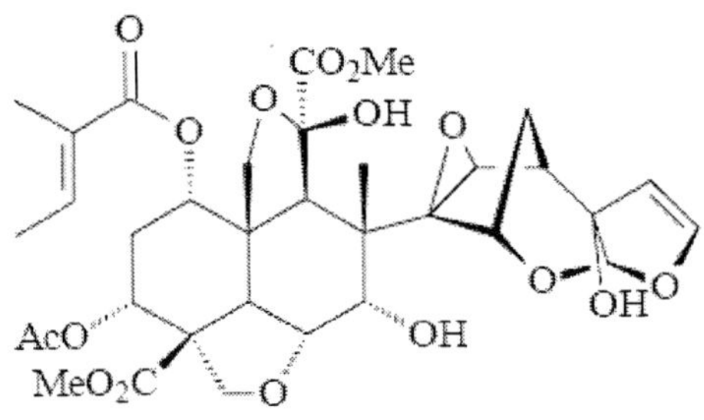

Azadirachtin

Powder from leaves and fruits of annona (*Annona cherimola*) (Figure 70) growing in tropical America is applied by local population to protect wheat against beetles *Rhizopertha dominica.* Annonine (reactant from *Annona cherimola*) is not practically used due to expensive isolation and purification.

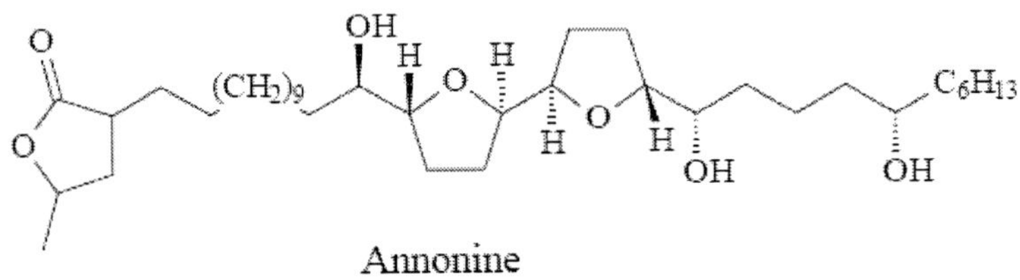

Annonine

Figure 70. Annona fruit (Annona cherimola).

Extract of fungi *Lactarius* containing sequiterpene acids possesses expressed antifeedant activity against pests of grain and grain products *T. confusum*, *S. granaries*, *R. dominica.* Activity of N-acylated derivatives of phenylisoserine – main reactant of fungus extract [169] is much higher.

Table 10 presents repellents officially allowed to be applied in the territory of the Russian Federation.

Table 10. The following repellents are allowed to be used

Preparation	Manufacturing country	Preparation form	Reactant	Leather working	Clothes treatment
Dipterol super spray	Croatia	spray	DEET 29%	yes	yes
Mosquitol anti-mite	France	spray	Alfamethrin 0.2% and DEET 7%	no	yes
Pretix	Russia	brick	Alfamethrin 0.45%	no	yes
Reftamid taiga	Russia	spray	Alfamethrin	no	no

Chapter 17

17. Insecticides of Microbiological Origin

Preparations obtained from spore bacteria *Bacillus thuringiensis* are used for insect control These bacteria produce two kinds of toxins one of which is toxic either for the mammal. To produce standardized insecticides selection of *Bacillus thuringiensis* is made, technologies for isolating and purification of endotoxins to get δ-toxin less toxic for the mammal are elaborated. It is proteic crystalline substance – parasporal body. When getting inside insect intestine the toxin is modified and reacts with intestine sides destroying it so that intestine content gets into hemolymph resulting in total sepsis. Metabolite of *Bacillus thuringiensis* endotoxins – dibeta (thuringiensin) – reveals strong insecticide properties [170].

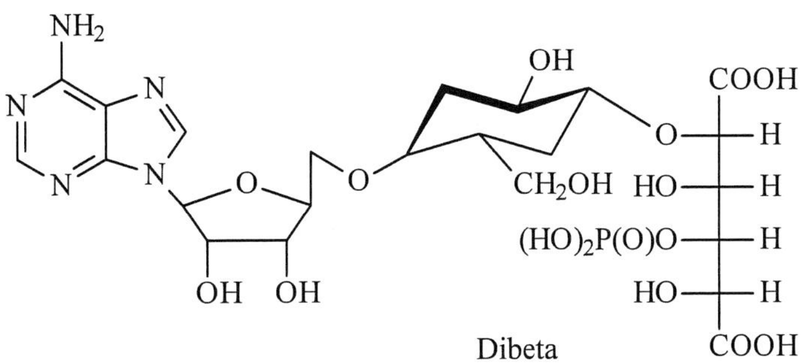

Dibeta

To control ectoparasites of farm animals macrolide antibiotics based on avermectines and ivermectines are applied. Synthetic analogs possess far more activity – even minor modification of natural avermectine structure by introducing methylamino group leads to increase of activity dozen times [171]. Preparation with trade name “Agravertin” as 0.2% emulsion concentrate is applied to control mandibulate and sucking insect-pests. Reactant is a complex of natural avermectines produced by geophilous fungus *Streptomyces avermitilis*.

The following toxic-based preparations produced by *Bacillus thuringiensis* – lepidocyde-50, dipel, bitoxybacilline are allowed to be used in Russia [172].

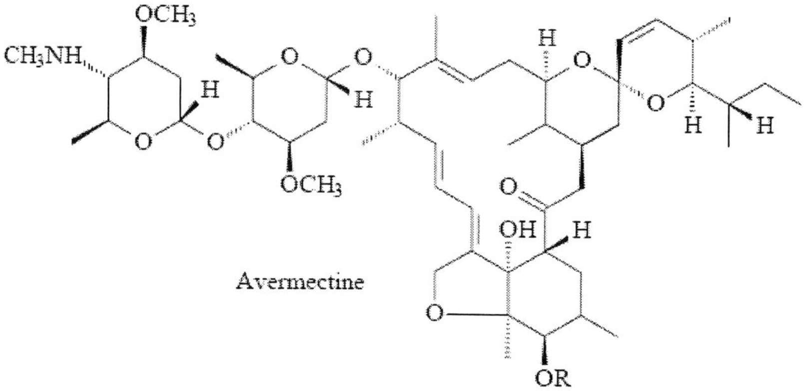

Substance slightly toxic for insects belonging to pyrrole derivatives – dioxapyrrolomicin is isolated from geophilous fungi *Streptomyces*.

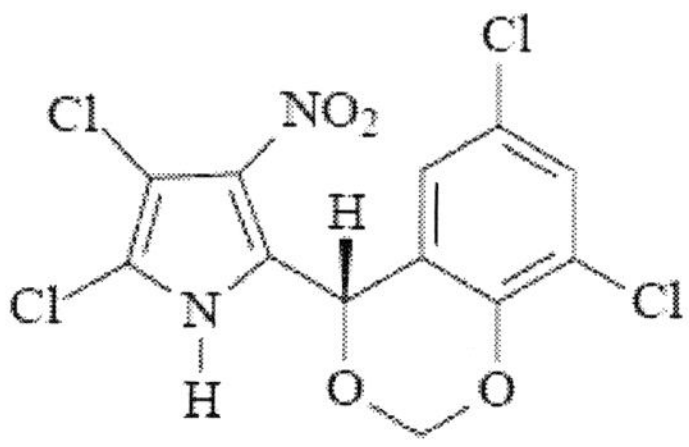

Dioxapyrrolomicin

Synthetic analogs of this substance containing electron-withdrawing substituents in position 3,4- or 2,4-benzene ring reveal much higher insecticide activity, for example preparation AC-303630 is active to 16 families and 5 orders of rodents [173].

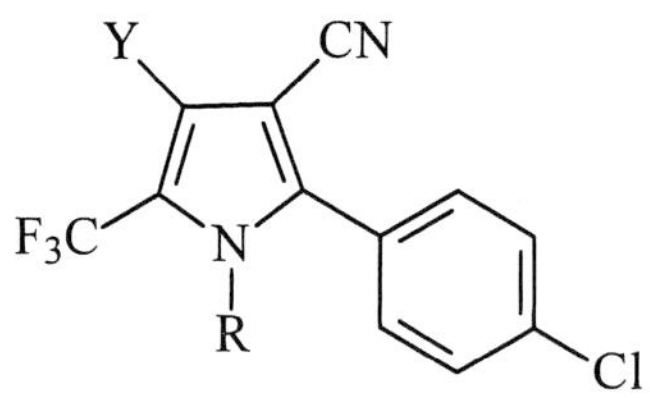

AC-303630

R = CH_2OEt, Y = Br(a), F(b); R = H, Y = Br (c)

Preparation AC-303630 is synthesized proceeding from N-(4-chlorphenyl) trifluoroacetamide (131). Treatment of the latter with phosphorus pentachloride with further cyclocondensation of the resultant chloramine 132 with 2-chloroacrylonitrile in dimethyl formamide results in 2-trifluoromethyl-5-(4-chlorphenyl)-4-cyanpyrrole (133), bromination of which over sodium methylate gives bromine pyrrole 134. Target preparation is obtained by alkylation of bromine pyrrole 134 with diethoxymethane.

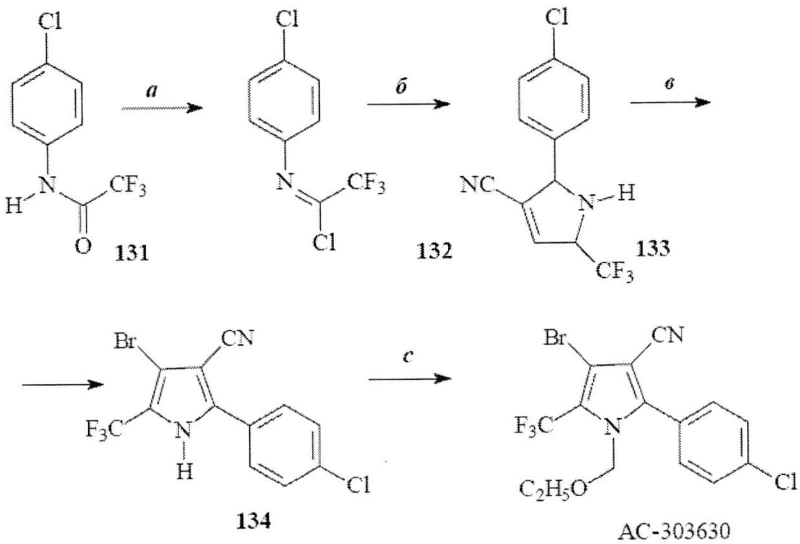

Reagents: a.PCl_5; б.$H_2C{=}C(Cl)CN$/DMF; в.Br_2/MeOH; г.$CH_2(OEt)_2$

Agravertin (complex of natural neurotoxins of avermectine series obtained as a result of microbiological synthesis from geophilous fungus *Streptomyces avermitilis*) affecting nervous system of insects, mites and nematodes causes paralysis in them, as a result the pest stops feeding, loses motion activity and dies soon.

Positive features of Agravertin are wide spectrum of phytophages being affected, small dose of consuming reactant, short safety intervals (2 days), allowing treatment to be made on fruit-bearing plants. Besides, pests do not become addictive to the preparation, it possesses soft environmentally-safe and hygiene and sanitary characteristics, with no bad and strong smell. The preparation can be applied on any green crop and cane fruits, as well as on ornamental plants under all technologies of growing.

After getting into the pest organism Avermectines activate synthesis of *gamma*-aminobutyric acid which is neurohormone of inhibition in nerve cells of invertebrates. Nervous system of vertebrates (including humans) is different, therefore, avermectines are actually harmless for humans and domestic animals. Due to this, avermectines are used in medicine and veterinary medicine, including injection forms to fight against endoparasites and dermal parasites.

In crop production avermectines (in particular, Agravertin - 0,2% emulsion concentrate) are actively used in fighting against Colorado beetle larvae, different aphid species, in extremely low doses it totally destroys populations of phytophagous mites and larvae of lepidoptera (totrices, seedworms, cutworms, etc.). To control gall eelworms another preparative form of Agravertin – wettable powder added to soil at plowing is used. For more than twenty years of applying avermectines abroad no resistance has been revealed in pests.

Agravertin is low toxic for warm-blooded, it possesses neither systemic nor fumigation effect (it neither penetrates plant tissues nor affects through respiratory system), and does not cause remote negative disorders. By parameters of food and skin toxicity it belongs to low-hazard substances. The preparation is applied during all period of plant vegetation, including harvesting period, as it quickly degrades in environment when affected by the sun and microorganisms. Agravertin is unlikely to get into groundwater, as its molecules are tightly bound with soil particles and penetrates soil for several centimeters only. For the same reason it is safe for beneficial soil organisms such as earthworms.

The preparation possesses selective action, i.e. it is less harmful for harmless and beneficial insects, which beneficially differs it from majority of modern chemicals. For example, coccinellids, reducing population of aphids, are actually resistant to the preparation. The same refers to predatory mites, ground beetles, ants and spiders. Selective ability of Agravertin is of great importance for ecological balance. Such entomophage as snouted mite phytoseiulus is not strongly affected by Agravertin treatments, while populations of phytophagous mites are totally suppressed. The same can be said about aphidius – agent for aphid fighting. It should be noted that this preparation is dangerous for predators and parasites when treated and under direct contact.

Important advantages of Agravertin are safety intervals – time to be passed after plant treatment till using it as a food. In majority of new pesticides this period lasts for weeks (20-30 days), while in Agravertin it is two days. It means that the preparation can be used to protect fruit-bearing plants, which is especially important for green crop and cane fruits (cucumbers, tomatoes, peppers, eggplants, currants, gooseberries, apple-trees, plum-trees, etc.).

Unlike majority of insecticides Agravertin can be applied in warm and dry weather when insect activity is the highest. At high humidity and low environment temperature it loses its activity and is quickly exposed to microbial decomposition.

Growing of transgenic plants producing endotoxins for protection against pests is quite promising direction. Introduction of *Cryl IA* gene into genetic apparatus of cultivated plants more than 100 times increases their injure resistance [174].

First generation of insect-resistant transgenic plants is based on using insecticidal albumen from *Bacillus thuringensis* (*Bt*). Second generation of transgenic plants obtained through using either *Bt*-albumens or other albumens with different spectra of activity against insect pests is being elaborated. *Bt* is known as pesticide for more than 40 years already. However, new era for growing insect-resistant transgenic plants has begun in 1981 only, when cloning and gene sequencing encoding these endotoxins were conducted.

Genes of δ-endotoxins are transplanted to plants through transformation of *Agrobacterium*. At present, there are transgenic dicotyledonous (tobacco, cotton, tomatoes, potatoes) and monocotyledonous (corn) plants with high enough level of endotoxins expression and good effect in field tests. It would be good, but it has turned out that some insects may have resistance to δ-endotoxins, and some insect species are absolutely resistant to them. Due to this a number of researches are being made to look for albumens of other nature. For example, polyphenol oxidases, proteinase inhibitors, α-amylases, and lectins. However, their transformation does not result in expected effect due to weak toxicity for insects.

New direction in searching for natural pesticides is screening of new potential insecticidal albumens. Objects of searching are plants, in particular, tropical plants with good insecticidal properties, media of microbial fermentation or albumens from available albumen library. To such new insecticides belong *Vip* albumens being metabolite products of *Bt* and cholesterol oxidase (ChO) from *Streptomyces* culture medium. ChO activity is compared with insecticidal activity of *Bt*-albumens, but it reveals toxicity to insects resistant to *Bt* toxins. All insecticides described in the article [175] of American authors are albumens. It is obvious, as while growing transgenic plants direct transgenation encoding these insecticidal albumens takes place. However, the authors do not consider possible toxicity and allergic properties of these albumens for humans. Almost all mentioned albumens affect on cells of insect intestinal epithelium causing "laking" of these cells. And such albumens like lectins cause agglutination* of erythrocytes. Problem of transferable insecticides affecting human health is sure to be considered while growing new transgenic plants.

Apart from introduction of genes causing formation of toxins in transgenic plant, another way to control insect-pests is possible. In the process of evolution plants were producing protective reactions to injures made by insect-pests. Due to stress under injure made to leaves or other part of a plant, it starts to produce albumen molecules, specifically affecting digestive proteolytic enzymes found in digestive tract of insects. Moreover, there is a trivial specific character by types of digestive enzymes dependent on co-evolution* of insect and plant considered as food [176]. For example, cysteine proteinase, L and B cathepsins, aspartyl proteinase of type D cathepsin are found in intestine of alfalfa weevil *Hypera postica* [177], while in lepidoptera order insects initial albuminolysis under gastrointestinal digestion takes place principally with proteinases of serine type [178, 179]. Proteinase inhibitors formed in plants are able to suppress (inhibit) activity of serine proteinases contained in insect digestive system which they need for normal digestion. Discovering of such effect made it possible to grow transgenic plants containing genes of proteinase inhibitors. Thus, transgenic plants of rice containing genes of cystatin I are noted for high content of inhibitors, and extract of this plant efficiently suppresses action of enzymes of papain, H cathepsin and proteinase of corn weevil (*Sitophilus zeamays* damaging rice grains [180].

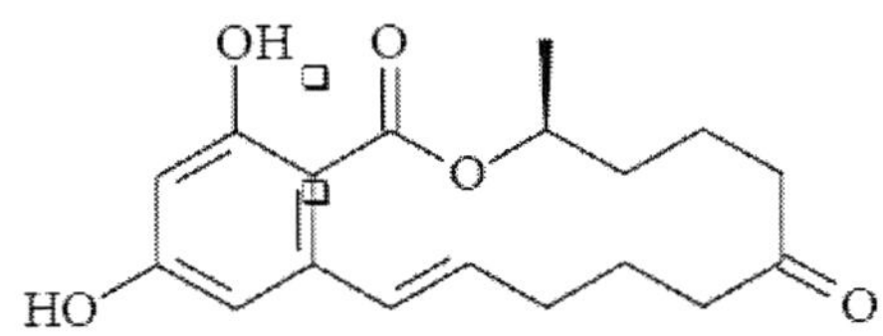

Zearolenone

Main problems with transgenic plants expansion are:

- unshared antigens will get into non-transgenic plants for account of transgenic pollen;
- transgenic plants intended to affect harmful insects only (e.g., Colorado beetle) will kill beneficial insects either (e.g., bees);
- transgenic plants may be allergenic or toxic ones.

* Agglutination (from Latin *agglutinare* – to glue) – the clumping together and precipitation from homogeneous suspension of bacteria, erythrocytes, etc. Takes place as a result of interaction between antibodies and antigens.

* Co-evolution (from Latin *co* – with, together + Latin *evolution* – expansion): parallel, mutual evolution.

- Sharp fertility drop in sow was noted on one farm of Iowa (USA). Apart from the fact that drop index was 80%, it was accompanied by "pseudopregnancy". Pigs looked like pregnant not only externally, but it was confirmed by tests and ultrasound investigations. However, after normal 113-days cycle all signs of pregnancy were gradually vanishing. And in 2-4 weeks the cycle was spontaneously repeated, throwing neither piglets nor slinks - fetus was completely "resolved" within the row. However, all problems disappeared when pigs were stopped to feed on genetically modified corn, where *Bt*-bacterium gene producing substances toxic for some pests was introduced. Distribution of *Bt*-corn was found to be accompanied by sharp growth of mold fungi colonies. This fungus produces a number of mycotoxins leading to pseudopregnancy. Decrease of corn sawfly population contributes to fusarium spreading, and exactly this pest is not affected by *Bt*-toxin. Pseudopregnancy is caused by zearalenone toxin produced by fusarium. Scientists from Iowa University making laboratory research of this case think that genetically modified corn is likely to be responsible for this. This case is supported by the fact that another 12 farmers from this state were facing similar problems, although they supported different technologies of pig-breeding, and applied to different veterinarians, were located in different Iowa counties, but all of them used genetically modified corn as food.
- Trypsin* in mosquitoes is to some extent a "hormone" simulating state of "go hungry" mosquito. Trypsin strengthening in adult mosquito female prompts of necessity to use all resources of organism to transmit nutrients to laid eggs. Therewith, only small amount of nutrients is used by the females and, accordingly, without food larvae either remain of the same size and die, or they will die in pupal stage when transforming into the mosquito. To make the mosquito larvae "feed on trypsin" the American researchers have cloned trypsin gene into yeast, which appears to be the best food for these insects. So, the Americans hope to decrease number of mosquitoes in some areas "by dispersing" transgenic yeast.

Entomophages** are the most important representatives of biological method and integrated plant protection. Among entomophages the most known are hymenoptera - ichneumon (Trichogramma, Aphelinus, Pseudaphycus) applied to control aphids, coccids and other insect pests, as well as some species of ants eating leaf-eating caterpillars; coleoptera - ladybirds, e.g., vedalia killing fluted scale, rove beetles eating green crop pests, some carabid beetles killing caterpillars; dipteran – tachina flies regulating growth not only of insects dangerous for farming, but some other invertebrates. Entomophages also include predatory bugs, mites; there are bacteria and viruses affecting insect pests, and fungi. Important role in nature is played by insect-eating mammals - Amphibia (frogs), some birds, as well as by mammals (hedgehogs, badgers, moles). Among entomophages used by humans insects rank the first: Hymenoptera (about 80), coleoptera (30), diptera (10), hemiptera (5), Neuroptera (3). There is single parasitism (development of one entomophage on a host), group parasitism

---

* Trypsin – Serine protease, enzyme from the pancreas of vertebrates.

** Entomophages (from Greek *entoma* - insects and *phagos* - predator) – vermin, parasites and other organisms dangerous for insects affecting natural regulation of their abundance.

(simultaneous development of several entomophages), superparasitism (development of many entomophages on a host), multiparasitism (development of many different entomophages on a host), as well as primary parasitism (affection of phytophages and predators), hyperparasitism (affection of other parasites) and kleptoparasitism (using of other parasites for searching and infestation of a host). Peculiarity of insect parasitism - entomophages – is in its lethality for a host, dissociation of stages for host searching and affection.

Dangerous predators found in granaries – true mites – feed on grain, cereals and contaminate with their excrements what is not eaten by them. Snouted mites like cheyletus (*Cheyletus eruditus*) are used to control them. First, true mite is cultivated on radish seeds. Then 20 thousand true mites and 150 cheyletuses are placed into 100 grams of lettuce. Under standard temperature and humidity the predator is propagated in special boxes so that in six weeks there will be not less than two thousand of them in 100 grams of nutrient medium. When true mites are fully eaten up, left cheyletuses are separated from the medium and placed into paper bags. At 2-5°C dry mites are stored for at least three months before being applied. Methyl salicylate with attractive properties can be applied for attraction of snouted mite *Phytoseiulus persimils* to mites damaging stored food [181].

Habrobracon *Habrobracon hebetor Say* is hymenopterous parasite (Figure 71) of caterpillars of cotton worm, European com borer, codling moth, flour moth and other (over 60 species) lepidopterous pests actively invading green crops, forage crops, fruit crops, cotton or stored grain, flour and dried fruit stored in warehouses. The insect parasitized on larvae of *P. interpunctella, E. kuehniella, C. cautella, E. elutella, G. mellonella, S. cerealella, Vitula edmandsii serratilineella.* Being developed in 2-4 generations these pests make significant damage and, therefore are regarded in all regions as objects of chemical control. Methods for applying this parasite against cotton worms and other worms on tomatoes have been made. Invasion of one hectare of tomato crops with 700-2000 species of habrobracon makes it possible to decrease crop losses and at the same time to reject 2-3 insecticide treatments. Seasonal norm is to be distributed considering development and propagation of pests (habrobracons are recommended to be put into crops twice within one generation). Habrobracons are placed diagonally in 5-6 points of the infested field. 0.2-6% of fruit is damaged under biological protection, while 2.8–11% is damaged under chemical protection. Application of ectoparasite *Anisopteromalus calandrae* allows population of maize weevil *S. oryzae* to be reduced by 58% [182].

Figure 71. Adult habrobracon.

Imported fire ants *Solenopsis invicta* being high adaptable and aggressive have become a great problem in the USA for pest control specialists. In Brazil – home of these ants in their nests *Beauveria bassiana* fungi were found, which are quite pathogenic for ants and safe for vertebrates. As soon as the insect contacts fungus spores, the latter stick to its chitin forming germ tube afterwards. Then the spore infests the insect by getting inside it.

The fungus has vegetative proliferation and it covers the whole insect body. After ant death the fungus leaves its body and proliferates in the environment. New spores are formed infesting other ants.

Traps with spore-infested bait attract foragers, and thereby promote to fungus propagation in ant nests. Even spore touching is enough for infestation.

Long incubation period makes possible to achieve high level of fungus proliferation. Maximum destruction of ants takes place on 3-4th day. Field tests have revealed that up to 50% of ant nests were destroyed by one treatment, with the population in the treated area being decreased by 60-80%. The preparation application inside premises gives better results. Microcapsules containing fungus spores make possible to increase efficiency of application even outside under higher humidity. Microcapsules sticking to ant limbs get inside the nest thereby infesting insects. Nematodes *Steinemema carpocapsae Weiser* are entomopathogenic organisms applied to control pests of different crops. *S.carpocapsae* affect large number of pests belonging to different insect orders (Figure 72).

Nematodes are applied depending on either culture and mode of pest existence, or by spraying plants (on green crops for controlling Colorado beetle and cutworms), or by introducing nematode suspension into soils at chemigation (to control codling moth in orchards, wireworms on tobacco and corn, etc.). Immature larvae of the III instar are usually used. Number of treatments with nematodes depends on insect population concentration.

Nematode amount depends on number of pests and usually is 30-90 per one caterpillar (larva) (i.e. 10 liters of preparation or 13.4 billion nematodes are required per 1 ha, at 10-13 thousand of species containing in 1 ml of suspension). While introducing nematodes at chemigation number of nematodes is to be 1.2-1.5 times increased.

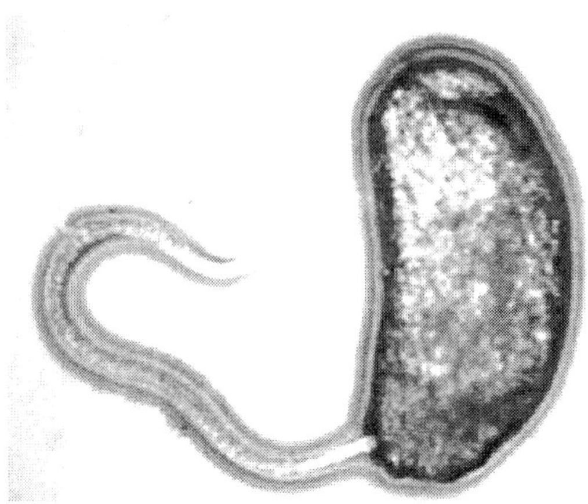

Figure 72. Nematode getting inside larva through rectal orifice.

Predacious fungi are applied to control nematodes. Fungi *Monacosporium cionopagum* are able to catch and keep crawling nematodes. By killing their victims they gradually digest them. Such peculiarity of predacious fungi feeding is stipulated by the fact that fungi inhabiting woody substrates lack nitrogen which is supplied by eating nematodes (Figure 73).

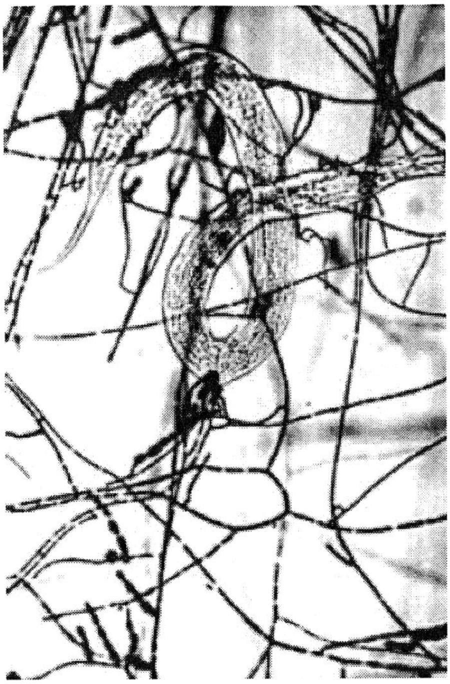

Figure 73. Fungus Monacosporium cionopagum, holding nematode.

Fungus mycelium possesses thin additional vegetative hyphae secreting toxin. This toxin immobilizes nematodes, while other hyphae enter inside the victim and digest it. Pleurotus toxin (ostreatin) also affects dew worms – large brandlings relating to earthworms. Main function of ostreatin is protection against mites, springtail and tardigrade feeding on fungus mycelium [183].

Chapter 18

18. Methods for Researching Insecticides Affecting Insects

Insecticides are usually analyzed as solutions of different concentrations in acetone, rarely in mineral oil. Standard insect used for analyzing toxicity is a housefly *Musca domestica* due to simplicity in growing and relatively short period of development. As LD_{50} toxicity is usually expressed in micrograms per insect weight, one must know average weight of the tested insect. Effective dose is introduced with calibrated syringe by regulated micrometer, giving one micro liter of insecticide solution. Acetone or another solvent is evaporating leaving insecticide on cuticle surface of the fly. To prevent flies from crawling away during the test adult flies can be immobilized with CO_2, however, as soon as insecticide is applied insects are to be placed into a cell with sucrose syrup for feeding them. Circadian rhythm is to be considered at testing. It is due to the fact that insecticide affection changes during the day. Morning tests significantly differ from that of in daytime. Different data can be obtained even within several days at the same time of the day.

To fix duration of insecticide action the substrate is moistened with tested solutions of certain concentration described above and dried. Insects are placed to the substrate for some time. Dry leaves of plants, slide plate or any other material can be used as substrate. Time for insecticide exposing is found experimentally until a number of lethal cases of insect death is reached within certain period of time.

Applying sprayers against insect swarms is not always leads to reliable results. Partially it is due to ability of insecticide to be crystallized and inability to penetrate through insect cuticle, more precise analyses are received when oil solution is applied.

Sometimes it is necessary to define toxicity by introduction of certain amount of insecticide. In this case injections of insecticide saline solution are made, which exactly correspond to ionic composition of insect hemolymph. Injections to adult flies are to be made into the upper part of thorax. As intrinsic pressure in adult flies is lower than atmospheric pressure, insecticide solution as micro drops will be quickly absorbed. The wound is pressurized with hemolymph, resulting in no significant mortality among tested insects. Injections to larvae are more difficult task due to absence of hard cuticle for introducing syringe needle. However, injections to caterpillars can be made through prospores which are hard enough to prevent outflow of inner contents.

Insecticides can be mixed with food. The flies readily consume sucrose syrups and thereby get the required dose of the preparation.

Initial bioscreening of juvenoids is made in laboratories on different test-objects, e.g. on pupae of yellow mealworm beetle *T. molitor*.

Oil solutions of appropriate substance in doses 100; 10; 1; 0.1 and 0.001 microgram/species is applied to covers of molting pupae. Results of treatment are evaluated by using ID_{50} – inhibiting dose of substance which allows 50% of intermediate forms to be received under standard terms of treatment (species with both signs of larvae and pupae, as well as intermediate forms between pupae and adult insects). Comprehensive tests of juvenoids which passed through initial bioscreening are made on laboratory populations of maize *S. oryzae*, granary *S. granarius* weevils, saw-toothed grain *O. surinamensis* and red grain *C. ferrugineus* weevils. Tests are made with beetles of different instars and including both male and female populations raised on wheat grain at 27 ± 1°C and 14% humidity.

Certain amount of tested preparation, considering reactant in it, is dissolved in certain amount of distilled water. Then 0.4 ml of the fixed concentration solution is taken with micropipette, and 40 g of grain is treated in Petri dishes. Control grain is treated with the same amount of distilled water. 30 beetles of this or that pest are placed onto treated grain approximately in 1 hour after treatment.

Then insects are put into thermostat at 27 ± 1°C and atmospheric humidity of 70±5%. Three weeks later mortality of beetles-parents is defined by taking them off the substrate, and then number of first generation brood is registered within 3 months. Tests are repeated three times. Influence of juvenile hormone analogs is defined by decrease in population in the tests as compared with the control. Hormonal effect is evaluated under 5-grade scale: absence of effect, i.e. formation of normal pupa (or imago) without visible morphogenetic abnormalities is 0, and maximum effect – secondary molt at the same developmental stage is 4. Other values mean different degree for keeping signs of previous developmental stage after molt to another one.

- Larvae of flour beetles, caterpillars of flour moth and other pests of grain can be the reason for dangerous disease - hymenolepiasis caused by rat tapeworm (*Hymenolepis diminuta*) which parasitizes mainly in rodent intestine, and rarely in humans. Infestation takes place at swallowing of parasite larvae together with their intermediate hosts – insect larvae.
- Damage caused by insects is defined not only by weight of damaged grain. Under strong infestation grain germinability is lowered, milling behavior is deteriorating, it is clogged, its temperature and humidity raise. Besides, nutritive value of grain and grain products is decreasing. Infestation with *T. molitor* changes aminoacid and mineral composition of stored products.
- It is found that with increase of flour beetle larvae number in grain, nutrient value of grain and that of meat-meal and bone flour made from animals fed with infested grain is decreasing. Content of main aminoacids such as methionine, cysteine, lysine, tryptophane, arginine, histidine, phenylalanine, as well as of calcium, phosphorus and sodium is also decreasing.

- Development of storage pests, esp. mites in stored grain affects taste and smell of grain. When their number is small the grain is of honey-like smell, further reproduction and activity of mites result in bad smell of rotten eggs (hydrogen sulfide).
- Corn-bug leaves very active proteolytic enzymes in grain. Strong wheat is regarded as weak one, when contained 3-4% of damaged grains. Gluten from bug-damaged grain affected by these enzymes is quickly rarefying. Baked bread is of small volume and porosity, solid and covered with small cracks, and tasteless.

Insect response to Sex pheromones is assessed either by behavioral response (in laboratory or in nature), or through laboratory electrophysiological tests applying special equipment.

Influence of exogenic ecdysteroids on insect growth and development is analyzed by different bioassays: feeding on phytoecdysteroids with food, making injections, as well as larvae dipping into solutions containing ecdysteroids. Action of ecdysteroids is assessed by number of indicators: survival rate, growth characteristics, changes during molting and pupation in larvae. Survival rate, fertility and viability of the brood are assessed in adult species. Antifeedant activity of phytoecdysteroids is also defined. Effect of action of exogenic phytoecdysteroids appeared to strongly depend on insect species and its developmental stage, applied phytoecdysteroid concentration, method of introduction into the insect body, as well as on chemical structure of the compound.

Behavioral responses in the laboratory are studied in the following ways:

- In open vessel (jar, glass, etc.) with living insects through which air stream with pheromones is passed.
- In small box with experimental insects which are affected either by air stream with tested substance or by filter paper soaked with this substance.
- In olfactometers* which may look like straight tubes, U-tube, round enclosed area, rotating circle, there are olfactometers equipped with feeders, as well as olfactometers connected to chromatographic system, as boxes-traps.

Response of flies to Sex pheromone was studied by a number of researcjers. Models of flies (pseudo flies) made from black laces were treated with extracts of virginal fly females and placed into Petri dishes among males. Males were attempting to mate with pseudo flies. By using olfactometer as large box into which glass traps with virginal fly females or extracts of them were placed, males got excited after air stream was passed through the trap onto them, and they were flying to the glass trap.

---

* Olfactometry (from Latin *olfaceo* — to smell and Greek *metron* — measure) - measuring olfactory acuity by special devices – olfactometers.

Chapter 19

19. Zoocides (Rodenticides, Raticides)

To control rodents greatly damaging different stored products zoocides* are applied. Apart from direct damaging foodstuffs the rodents make more damage by contaminating them with their biowaste and infecting products with different harmful organisms. Sea onion *Urginea maritima* (Figure 74) containing scylliraside was applying still in the Middle Ages.

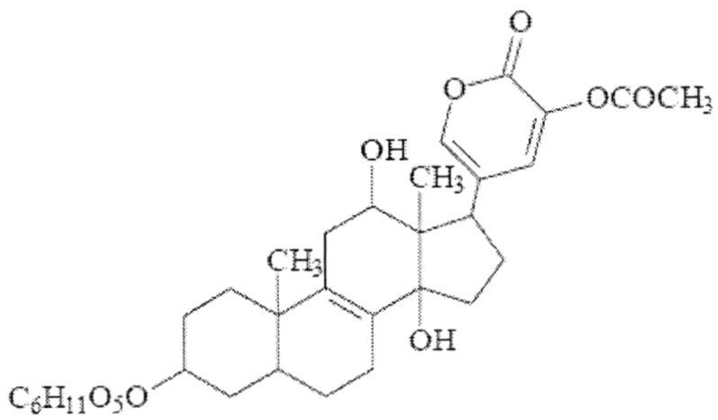

Scylliraside (LD_{50} 0,.4-0.7 mg/kg)

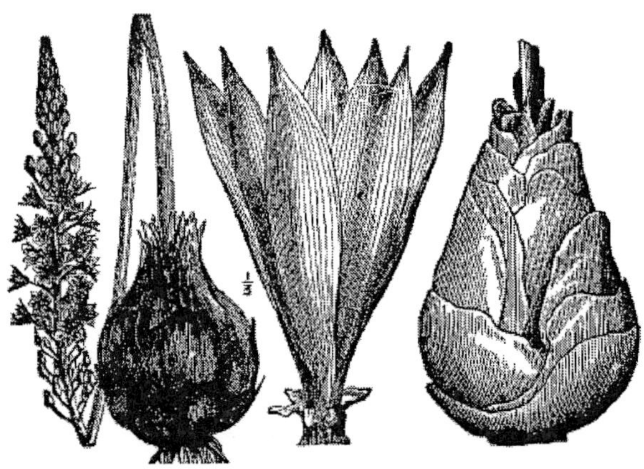

Figure 74. Sea onion Urginea maritima.

---

* Zoocide (from Greek *zoon* – animal and Latin *caedo* – kill) – means to kill vertebrate pests.

Coumarin derivatives possess anticoagulant properties. They do not cause quick death of the animal and therefore can be applied against such shy animals as rats. There are exploratory active individuals in each rat population, and it is them who are the first to test different products and baits. If some time after testing the new food this individual dies, the whole population rejects this new food.

Therefore, it is impossible to apply immediate action toxins against these rodents. From this point of view, coumarins are ideal substance for fighting, as their action is prolonged and the rats cannot associate death of one of the individuals with feeding certain bait.

- 4-oxycoumarins and dicumarin contained in white clover and yellow sweet clover prevent blood from coagulation. Cattle feeding on these plants often suffers from bleeding sickness and sometimes dies. If only 0.0026% of dicumarin is contained in dry clover hay, it may be enough for cattle loss. Anticoagulative properties of dicumarin and its derivatives are widely applied in clinics for treating and prophylaxis of trombosis.
- Hemorrhage in rats can be provoked by introducing oxycoumarin in dose of 3 mg/kg only. Coumarin is a drug for rabbits, sedative and soporific for dogs, horses, mice and other animals.

Systematic feeding of such preparations results in internal hemorrhage in rats with further death. One of such preparations is warfarin – 4-hydroxy-3-(3-oxo-1-phenylbutyl)coumarin, it is synthesized by condensation of 4- hydroxycoumarin with benzylidene-acetone (136):

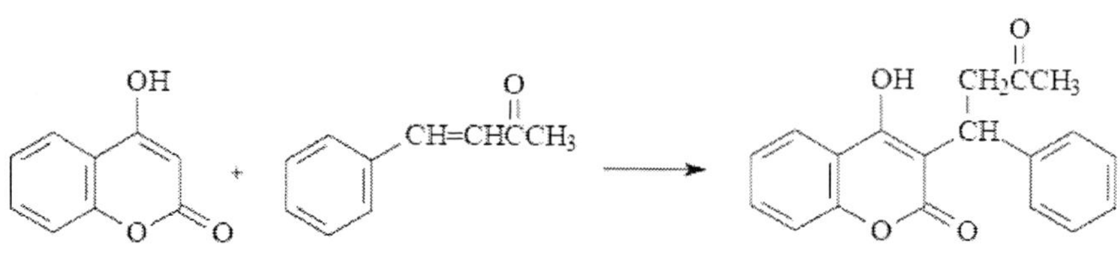

135 136 Warfarin

Warfarin activity is increased if applied together with vitamin D_2, which is zoocide of chronic action damaging liver and kidneys. Coumachlor - (4-hydroxy-3-[3-oxo-1-(4-chlorophenyl)]coumarin is more active derivative of coumarin:

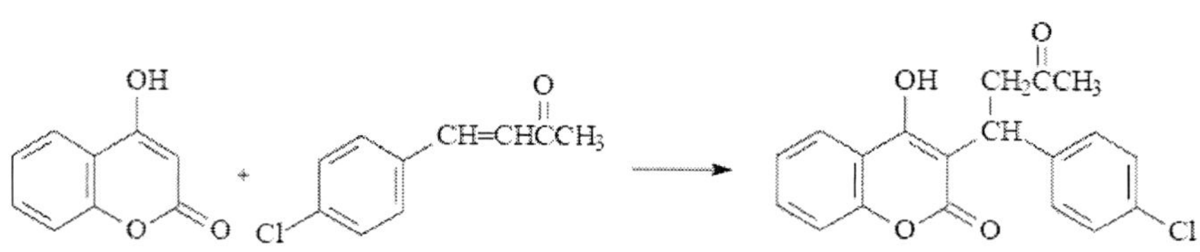

Coumachlor

- Furocoumarins (bergapten, xantotoxin, imperatorin, psoralen, etc.) and vegetable extracts containing them (bergamot oil) possess photosensibilizing property, as they increase unprotected skin exposure to UV irradiation.
- Main mechanism of anticoagulant action of coumarins is their antagonism in relation to vitamins K, which are necessary components in synthesis of factors for blood coagulation. Warfarin and other phenol anticoagulants inhibit synthesis of those factors of coagulation, which need vitamins K for being active. In this case coumarins behave like structural analogs, antimetabolites of vitamins K, displacing them from the bond with protein part of appropriate enzyme system [184].

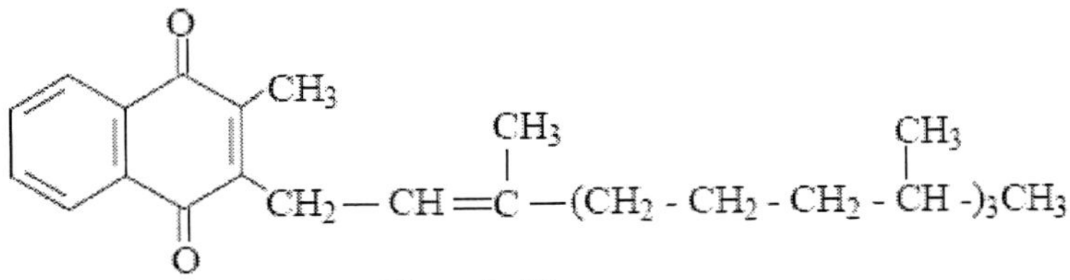

Vitamin K$_1$

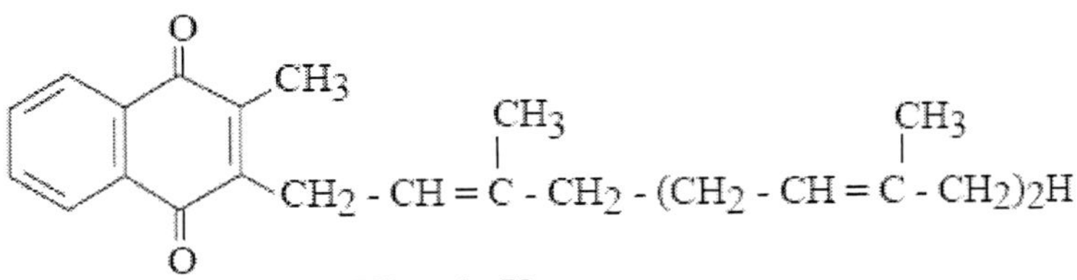

Vitamin K$_2$

Bacteria *Salmonella enteitidis var. Issathenko* (bacterodencide preparation) are applied for rodent control, causing in rodents the disease similar to typhoid and not dangerous for humans and domestic animals.

Acute poisons are thallium sulfate $TlSO_4$, barium carbonate $BaCO_3$, zinc phosphide Zn_3P_2 (LD_{50} 15–30 mg/kg), sodium fluoroacetate FCH_2COONa (LD_{50} 0.22 mg/kg), fluoroacetamide (fussol, fluoroacyl 100) FCH_2CONH_2. Zinc phosphide toxicity is likely to be due to isolation of poisonous phosphine within the animal stomach influenced by hydrochloric acid. This phosphine is as toxic as hydrogen cyanide:

$Zn_3P_2 + 6HCl = 3ZnCl_2 + 2PH_3$.

Table 11. Rodenticides allowed to be applied in the territory of Russia

Reactant	Concentrates (g – gels and pastes; l - liquid; p - powders)	Prepared baits
I generation anticoagulants		
Warfarin	Gelcum (g); Deccum (p); Warfarin-powder (p); Warfarin painted (p); Rodent ML (p); Rodent (p)	
Coumatetralyl	Racumin-powder (p)	*Cereals:* Racumin - bait
Difenacin	Ratindan (p); Ratindan-0.5 (p); Ratindan -1 (p); Ratindan -2 (p); Hedgehog (l)	*Cereals:* Zeracid *Bricks:* Difa

Table 11. (Continued)

Reactant	Concentrates (g – gels and pastes; l - liquid; p - powders)	Prepared baits
Ethylfenacin	Hilcin (g); Ethyphemacin-paste-2 (g)	*Cereals:* Barrier-Э.1у; Barrier-Э.1к; Barrier-Э.3у; Barrier-Э.3к; Zernocin-U; Zernocin-K; Zernocin-A; Polytox-bait; Raptor-sanitar (against rats) *Bricks:* Zernocin-brick-K; Zernocin-brick-U; Efa
Chlorfacinon	Brom-HF (l); Caid (l); Caid-powder (p)	*Cereals:* Caid; Rodefacin-U, Rodefacin-K
II generation anticoagulants		
Izoindan	Holdane (g) Indan-fluid (l)	*Cereals:* Fumitox-RAT; *Granules:* Grindan K; Grindan U *Bricks:* Zerdan-block-K; Zerdan-block-M
Bromadiolon	Brom-BD (l); Lanirat-concentrate (l)	*Cereals:* Brom-topichida; Globol-Rodenticide; Desars-G; Desars-P; Lanitrat-bait; Rodialon *Granules:* Rott killer super. *Bricks:* Poteke; Rodentan-B
Difenacum		*Cereals:* Roban-grain bait
Brodifacum		*Cereals:* Vertox – grain bait; Rodefacum *Granules:* Vertox-pelletes; Clerat
Flocoumafen		*Bricks:* Storm or Cyclone
Difetialon		*Bricks:* baraki – paraffin blocks
Acute poisons		
Crycid	Crycid (p)	
Zinc phosphide	Zinc phosphide techn. 80% (p)	
Aminostigmin		*Cereals:* Amus
Vitamins "D"		*Cereals:* Raptor-sanitar

Chapter 20

20. Physical Methods of Control

20.1. Ultrasonic Repellents

Ultrasonic repellents use principle for affecting on rodent perception with high-frequency sound vibrations. Rodents perceive ultrasound of certain band both by organs of hearing and by the animal body surface. Cardiovascular activity is disturbed thereby, and distinct phase-wise changes in the nervous system expressed in depression of its functionality take place. Mention should be made of specific and individual sensibility of the rodents to ultrasound. Therefore, reaction can be different: from increment in motion activity during first 5-10 days to convulsive state. Animal behavior within the device action depends on distance and availability of objects absorbing ultrasound (soft surfaces, carpets, and wooden or paper containers in warehouses). Method of ultrasonic action cannot guarantee 100% positive effect due to:

- disturbance of reflex action in some animals resulting from consequences of diseases;
- obsolescent perceptive organs and mutation consequences.

Anyway, advantages of this method are that normal life activity of animals is disturbed if exposed for a long period of time, thus leading to bottleneck phenomenon in the monitored premises.

- SK Telecom Co., South Korea's top mobile phone operator, is expanding a new service allowing its customers to repel mosquitoes within a range of 1 meter using their cell phones. The researchers have discovered that certain sounds repel blood-sucking insects. Records with these sounds are reproduced as melody for cell phones. Subscribers can pay 3,000 won (US$2.55) to download a sound that SK Telecom says is on a frequency that annoys mosquitoes. Antimosquito device for SKT was made by Korean company Dots Mobile specializing in content deliveries for wireless networks. Sung In-jae for Dots Mobile, a Korean expert says only female mosquitoes require a blood meal and bite warm or cold-blooded animals. Male mosquitoes do not bite, but feed on the nectar of flowers or other suitable sugar source. Bloodthirsty female mosquitoes tend to avoid their non-biting male counterparts. According to SKT, the sound should be capable of clearing the insects within a range of one meter. The sound plays constantly and is, if at all, faintly audible to the human ear. This way of repelling the bloodsuckers will use up to 30 percent more of the phone's battery power. Lee Won-ja, a mosquito expert at South Korea's National Institute of Health, said the idea was good in principle but she doubted it would work in practice. "There are 54 kinds of mosquito in Korea, and they all make different sounds, so I don't think a mobile phone will prevent people being bitten," she said. Lee Won-ja thinks it possible that the mosquito-like sound could be audible to humans and prove irritating for them. Anyway, Mr. Sung was confident that at least 90,000, or 1 percent, of SK Telecom's wireless Internet customers would download the service.

LR-001 device (Figure 75) is intended for using inside premises. It repels the mosquitoes with high-frequency sound waves. Depending on the model they can be equipped with "on" sensor and be used as night light: 220-240 V, 50 Hz. Functions: single-frequency sound wave. Possibilities: automatic photosensor.

Figure 75. Ultrasonic pest repellent LR-001.

20.2. Insect-Trap Light

Using of electric-discharge insecticidal lamps in production premises revealed their ineffectiveness, as meshes of electric-discharge grid were intended for certain size (fraction) insects. Due to this, those premises where different size insects can be seen, e.g. cheese fly, house fly, and flesh flies are to be equipped with three insecticidal lamps with different mesh sizes.

Besides, these lamps are not applicable in moist and dusty premises, and remains of chitin and wings of burnt insects can be transferred by air to clean surfaces, on foodstuffs and humans. Industrial killer of flying day-time and nocturnal insects (models SQ-22, SCH-15,

Figure 76) is using in its work UV source with wavelength of 570 nm. It is absolutely for people's health.

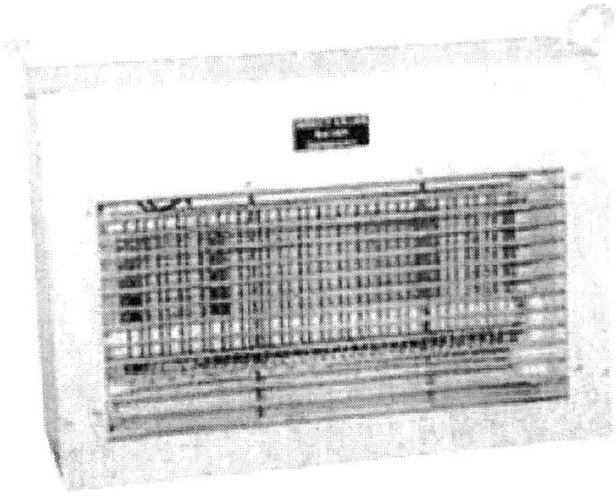

Figure 76. Insect-trap light SQ-20.

The device is connected to AC power. UV light source is switched on, thereby attracting insects. Flying up to the tube the insects touch energized metal grid and die. Higher power models are more effective, as they attract insects stronger.

It is used for killing insects in industrial premises and offices, in stores and warehouses of both raw materials and finished goods, including premises, where foodstuffs are stored (restaurants, etc.).

Voltage: 4000B. Effective area of action: 120 m2.

20.3. Controlled Gas Disinfection

Carbon dioxide or nitrogen replacing oxygen necessary for normal breezing of insects in closed premises is used to kill insects. Under appropriate exposure magnitude of insect population of different instars is significantly decreasing. This measure is especially effective if combined with temperature rise for up to 60°C in premises, with affecting no grain quality and its germinating capacity.

Until recently premises fumigation (Figure 77) was regarded as main way to control pyralids in bakeries. Traditionally once a year mills and other grain-processing plants are disinfected by methyl bromide fumigation. This method is expensive and labor-intensive one. The mill is to be preliminary sealed which is not always can be done. Paper covering of window crossing has actually no sense, as the gas easily penetrates through it outdoors. Besides, heavy methyl bromide immediately after releasing from tanks is settling down and, as a rule, living insects are left on upper 2-3 floors and in flour bins. During methyl bromide fumigation all activities at 30-50 m from the object to be treated are stopped, and people are not allowed to enter the premises for several days period of exposure and degassing. This area is to be guarded and a doctor should be present due to possible poisoning.

Phosphine fumigation is also allowed to be made according to instruction for control of grain products pests. This preparation is more preferable for mill disinfection due to its properties, in particular, specific smell, good distribution around the premises because of its density similar to air density, greater toxicity for insects, ease and less danger at work, and simple degassing.

Figure 77. Warehouse fumigation.

Disadvantages of fumigation method are short duration of its action. The mill is not protected against harmful insects when infected grain gets inside. Usually pest population is restored in 3-4 months; as a result they stay in the mill for almost a year causing unfavorable sanitary situation, destroying and damaging flour, impairing operation of drifts, aspiration and other equipment, threatening to infestation of the finished products. Fumigation is made by smoke pellets in closed premises (warehouses, storage bunkers, storage silos, etc.). Pyrotechnical mixture added by insecticide (fufanon, actellic, carbophos and others) or fungicide (thiabendazol and similar to it) forms the basis for smoke pellets. When insecticide is added to the pellet, it is used for disinfection of empty warehouses against flying insects – meal moth, Indian meal moth, flour moth, European grain moth, corn moth. When fungicide is added to the pellet, it is used against fusarium, phoma, osporos, silver scurf, gangrene of potato and dry rot.

20.4. Glue Traps

Glue traps (Figure 78) containing pheromone are widely applied to control number of harmful insects. Continuous monitoring allows critical threshold of insect pest number to be fixed, and to define the required doses, number and duration of treatment. Insect pheromone trap is a very sensitive means for detecting insects. These traps are simple in making, cheap and just several milligrams of reactant are enough for the trap to operate for a long time.

Due to high species-specific and selectivity of action, special training of personnel is not required for identifying species of entrapped insects. High pheromone sensitivity of insects makes pheromone traps ideal means for detecting certain insect species on large areas, including quarantine pests in customs stations. However, it is difficult to interpret data by larva number which make main damage in storehouses for food and raw materials, because reliable correlation between number of trapped insects and population number on the whole is hard to define. Disadvantages of pheromone traps are their orientation on male species only attracted by sexual attractants of females.

Figure 78. Pheromone trap.

But along with decrease of insect pest population by reducing the number of males, number of virgin females laying no eggs is increased. Absolute absence of pheromone toxicity for warm-blooded animals and humans, minimum residual pheromone content in products make them ideal means in the system for integrated protection of food stocks and foodstuffs against insect pests, including biological methods with usage of predators and parasites.

Synthetic attractant is applied onto special preparative form (different sorts of rubber, of red color mainly, polyethylene ampoules or rods soldered on both sides made of different polymers), being attractive within 2 months. In other cases pheromone is introduced directly to the glue. Before being used the pheromone capsules are kept in fridges in foil packages or glass vessels with tight plugs. When applied the pheromone capsules are taken by clean hands, as insects are very sensible to foreign smell.

Traps of three types are made to control pyralids: closed with glue surface, glue tapes and live traps. They are intended both to reveal pestholes and their number and for mass capture of males. Set of closed glue trap ("Attracon" type) consists of cardboard frame, pheromone evaporator (dispenser) and glue leaves to be replaced after glue surface is filled with insects. Glue tape is a band of laminated polyethylene paper with one-sided glue-pheromone covering. Such traps are thrown out after being filled with trapped butterflies. Plastic live traps consist of a frame, pheromone evaporator and container for insect capture.

Each of these traps has both advantages and disadvantages. Cardboard traps "Attracon" and bands are cheaper and capture well all pyralid species. However, if insect population is great either glue leaves or traps (trap-bands) are to be often replaced. Besides, they are not effective in premises with strong flour dust. Live traps do not require special maintenance (except for periodic dispenser changing). However, they are more expensive and reportedly less effective against almond and cacao moths.

Quite often only pest imagoes can be found in storehouses and similar premises. Sometimes, larvae are difficult to detect due to their small number, continuous production process, and their storing in different packages. At the same time continuous turnover of raw materials and production is always threatened by pests.

To detect pyralids one trap is placed to cover 700–1000 m^3 (reaching 10°C and higher in non-heated premises). Further the following measures can be taken: when local pestholes are revealed – visual assessing, cleaning of available mechanisms and equipment and making local chemical treatment; under general infestation – safety precautions are accompanied by mass capture of males through pheromone traps (if there are 3 butterflies per 1 trap in a day). If number of trapped butterflies is increased (over 15 per one trap in a day within 10-days period) chemical treatment is to be made with prior manual cleaning of available mechanisms and equipment.

Traps of any kind are placed each 125-200 m^3 in premises at 1.5 m above floor level for mass capture. Theses traps are to be located at least 2 m from working mechanisms (equipment vibration decreases capture efficiency) and windows. Traps are not recommended to be placed where flour dust is available, thus decreasing quality of glue surfaces [185].

Traps are to be distributed around the premises depending on real conditions and considering possible places for pest development and localization.

Trapped butterflies are recommended to be checked every 7-10 days. When using glue traps enclosed leaves are to be replaced periodically. Generally, enclosed leaves are replaced when 3/4 of glue surface is covered with butterflies, pheromone dispensers are replaced every 2 months, trap frames – when worn.

Efficiency of mass capture depends on density of insect population. If it is high initially, prior chemical treatment is recommended. Anyway, regular general preventive measures are necessary condition for effective usage of pheromone traps against pyralids. Removal of spilled flour in premises and periodic cleaning of mechanisms and equipment reduces hidden pestholes. More often, if possible, opening of hatches promotes to butterfly catching gathered inside the mechanisms. Failure to place traps as recommended, as well as their wrong location, negatively result in efficiency of mass capture. The same can be said about non-observance in periodic changing of dispensers and glue traps.

To catch flies Sex pheromone of the house fly - muscalure (muscamon) is used in the composition of insecticidal means: poisonous food baits, glue tapes and other glue surfaces. “Catch-master” series preparations consisting of glue applied to the cover and Sex pheromone: glue traps for flies “Gold Stick” and “Small Gold Stick”, glue traps for flies as leaves are registered, manufactured and used in the USA.

The following baits containing carbamate methomyl (1%) and Sex pheromone of flies are registered and applied in Russia: 0,025% of muscamon in “Gold Marlyn” (“Zoecon”, USA; “Sandoz”, Switzerland), 0,0125% of muscamon in “Moscacid” (“Zapi”, Italy) and 0,251% of muscalure in “Bait” (“Bayer AG”, Germany).

In 1973 French scientists made biconical bait successfully applied in many regions of Africa against riverine species of tsetse flies. The trap is made of material and is easily soaked with insecticide. Riverine species of tsetse inhabit narrow strips of riverside vegetation and fly along the rivers. Therefore, they often get into traps placed along the river banks. However, effect from their application was quite low. Effectiveness was increased significantly after using attractants simulating bull's trail, i.e. target of tsetse flies. Using of microelectrodes installed into fly antennae made it possible to define that these insects are attracted mainly by carbon dioxide, acetone and 1-octene triol. These data were used for making new traps. The trap looked like dark rectangle made of cloth backlined with thin net invisible for flies. Both cloth and net were soaked with deltametrine insecticide, and dark rectangle – with mixture of acetone and attractant. Insects attracted by smells and sight of dark rectangle get on cloth and were caught in net. In all these cases they got lethal dose of insecticide. Such traps were successfully applied in Zimbabwe, Malawi and Zambia, resulting in total elimination of tsetse in the island in Carib lake. Using of four traps per one square kilometer in total area of 600,000 sq.km in Northern Zimbabwe totally eradicated flies.

20.5. Heat Treatment

One of preventive measures for pest control is grain cooling. It is made by two steps: up to 20°C, to drastically reduce insect development, and to the temperature below lower temperature threshold of insect development. Such temperature values are given in Table 12.

Table 12. Lower temperature threshold of pest development

Pest	Lifetime by most stable stages (days) at			
	0°C	-5°C	-10°C	-15°C
Granary weevil	67	26	14	19 h
Maize weevil	17	12	4	7,5 h
Confused flour beetle	12	5	5	5 h
Dark flour beetle	19	5*	2	4 h **
Red grain beetle	112	32	20	24 h
Saw-toothed grain beetle	22	13	3	24 h
Lesser grain borer	17	10	1	7 h
White-marked spider beetle	219	164	36	17
Pea beetle	>404	>260	130	6
Corn moth	25	9	2	2
Flour moth	116	24	11	2
True mite: feeding stages	486	18	7	1
True mite: eggs	386	168	57	1
Elongated mite	85	24	21	1
Haired mite:				
Feeding stages	50	18	8	3
Hypopi	>500	>500	330	124

* At – 2.5°C.
**At – 18°C.

There is no increase in the number of insects in grain cooled till lower temperatures, and they soon die. Under such conditions grain is allowed to be stored without disinsection. Grain is cooled either inactively i.e. aeration of premises, or actively through stationary or mobile ventilating units, transmitting through grain separators or conveyers, cooling or drying chambers of grain dries, to be cold aired.

Grain cooling can be the main action to protect grain from insects in northern regions of the country, where weather conditions permit doing so.

In southern regions cooling is hard to be done due to high atmospheric temperature, therefore, other measures are taken there, e.g. grain treating with contact insecticides. It protects grain from infestation for several months.

After the grain is prepared for storage and loading into the granary it is to be regularly controlled for pests.

University specialists from Manitoba province (Canada) developed highly-effective method of wheat grain disinsection. This method is based on heat exchange while mixing infested wheat grain with the heat-transfer agent (hot silica sand). Mass of grain is recommended to be mixed with sand heated till 82°C, in the ration of 1:33. In 1.5 minute temperature of grain reach 50°C, under which all main stored grain pests die in 15 seconds: red flour beetle, granary weevil, red flour and saw-toothed grain beetles, and others. Grain is treated in electrified unit with large metal containers. All operations are automated. Silica sand (0.59-1.35 mm particle size) is heated in the first container. It is mixed with infested grain by simultaneous loading into the second rotating container specially isolated from heat releasing. Then the mixture is unloaded via the sieve, though which the sand passes and returns again to the first container for repeated heating, with the grain going to the granary. Thermal disinsection also suppresses development of molds *Alternaria alternata, Aspergillus* and others without reducing quality of bread and seed grain. As compared with traditional chemical treatment this method is more economic, available and does not contaminate products and environment with pesticides. Violating drying regimes leads to formation of burning or smoky smells.

Chapter 21

21. Insecticide Resistance in Arthropoda

One of main factors reducing effectiveness of destructive measures is arthropoda resistance to insecticides applied against them. At present there are over 500 arthropoda species making resistant populations [186].

For example, among 33 lines of maize weevils *S. oryzae* from nine regions of Pakistan 13 were malathion-resistant, with resistance being higher in public storehouses. Among 40 lines of confused flour beetle *T. castaneum* no malathion-resistant species were found in 11 regions. Phosphine-resistance was found in 5 lines [187].

Individual sensibility of the arthropoda to insecticides is different [188]. In natural arthropoda populations there are some individuals possessing natural reduced insecticide sensibility and survive when affected with doses lethal for other species. If lifetime of one generation is relatively short, and treatments with the same preparations are often repeated, surviving resistant species quite quickly increase in number and thus become main species of the population. Resistance development is also assisted by high insecticide dosage. In such cases insecticide acts as strong selective factor. Majority of natural populations of each species possesses enough genetic diversities for resistance to any insecticides, with the process itself and rate thereof being defined by peculiarities of arthropoda ecology, character of selection pressure (permanent, alternate, affected on one or all stages of development, etc.), by genetic kind of population.

There are natural (conventional) and acquired resistance.

Resistance is based on biological and biochemical peculiarities of organism. Sensibility level of individuals to poisons varies greatly even within one species. In number of cases females, older instar larvae, wintering or wintered individuals (eat less, with enough fat) are less sensible to the preparations. Acquired resistance is ability of individuals of certain species to survive and reproduce, affected by chemical, biological and other substances formerly suppressing their life activity.

Mechanism of insecticide resistance is complex and specific in each case. Resistance may have morphological, behavioral and biochemical bases. In first case individuals develop or are selected after pesticide affection, whose morphological structures prevent from pesticide getting inside the organism, e.g. chitin thickening, strengthening of cuticle are noted in resistant flies, due to which less poison gets inside the organism.

Under behavioral resistance the arthropoda avoids long contact with treated surfaces (irritability effect). They ignore insecticide baits (some species of cockroaches, flies).

The arthropoda possess enzymatic systems capable to neutralize majority of insecticides. These systems are likely to be expressed stronger in individuals with higher natural resistance than in sensible ones. Rate of insecticide excretion from the organism is increased.

The most important mechanism of resistance formation is obtusion of nervous system receptors thus resulting in loss of insecticide efficiency.

If insecticide affection is to stop, isolated laboratory population (on the whole) becomes more toxic-resistant after several generations. If the same preparation is applied again the resistance appears again and faster.

Race formed under affection of one insecticide can be cross-resistant to related and even non-related compounds. So, insects resistant to organophosphates often become resistant to carbamates, because mode of action of these groups is alike – they inhibit acetylcholinesterase. Organochlorines resistance is often accompanied by resistance (to a lesser extent) to synthetic pyrethroids, as despite differences in chemical composition these groups of compounds possess similar mode of action.

Chromosomes and genes responsible for resistance are studied and found in some insects and other arthropoda. Mechanisms of different resistance formation are widely analyzed depending on different factors.

Mapping on the drosophila genome made it possible to reveal DDT resistance gene (DDT-R). Enzyme of cytocromes P450 normally available in all living things decomposes different toxics, in mutated drosophila flies this enzyme is produced in such amounts that DDT decomposes before reaching vital organs of insects. Composition of other genes responsible for resistance in case of mutation is revealed (kdr, AChE).

Insecticide resistance of different species is forming later under the same conditions. For example, initial data of house flies resistance to DDT and hexachlorocyclohexane were received 6 years after these preparations application, and 4 years later to organophosphates. At the same time the resistance was forming slower in populations of little house flies, bluebottle flies and green blowflies, which means that these species differ by their genetic potentials. Despite application of chemical insecticides for many years, tsetse flies were unable to make resistance to applied preparations.

Resistance appears not only to chemical compounds. The arthropoda appear to resists also to biological agents, chemosterilants, and development regulators. Review of current resistance development shows that actually all arthropoda of epidemiological importance have acquired it, however, its level in different species is different. At present resistance to organochlorines, organophosphates and pyrethroids is revealed in flies, cockroaches, fleas and bed bugs, to organochlorines and organophosphates in lice.

Resistance development is often accompanied by decreasing of biological potential of population – lifetime of resistant individuals is decreasing, females lay less eggs.

Increase of insecticide dosage, application of preparation of prolonged action in small areas against both larvae and mosquito imagoes result in fast selection and promote to resistance acceleration. In number of countries measures to fight malaria have led to resistant population of mosquitoes of *Anopheles* genus. There are data on organophosphate-resistant mosquitoes of *Culex* genus.

Cases of arthropoda poliresistance (resistance to many substances) are often due to intensive application of many mixtures of different compounds in agriculture, which are often air-sprayed against pests of different crops (rice, cotton, etc.).

Process of resistance development consists of 3 periods:

- tolerance* period – when level of arthropoda sensibility to insecticides is decreasing 2-3 times;
- period of intermittent growth – when each increase of preparation dosage is reacted by the population with resistance increase;
- maximum level period – when program of measures to be taken is to be corrected. In these cases preparation changing or alteration is to be thoroughly planned, for their efficiency being not reduced.

Insect population can be called resistant if this phenomenon is confirmed by standard laboratory tests.

Each insect species has different resistance level when their control by any preparations is ineffective. Survival of at least 20% of tested individuals affected by certain dosage (concentrations) of applied insecticides is usually regarded as practical criteria for resistance.

One of important part of disinsection is early diagnostics of resistance starting and growing. Resistance monitoring is recommended to be made among economically significant species of the arthropoda in different territories, for which purpose trained qualified personnel, elaborated procedures and tests for resistance level determination, as well as criteria for evaluation of obtained results are required. Besides, new substances to be applied as insecticides are to be tested at early stage of their development to reveal possibility of cross-resistance to them. It is advisable to use compounds with narrow spectrum of cross-resistance.

To reveal resistance of malaria mosquitoes to insecticides standard test-kits of the World Health Organization (WHO) are used. However, this method is not the only criterion and in field conditions one should also take into account mosquito number, their physiological age and other indicators, which characterize insect population in the aggregate.

The following procedures are proposed to define level of sensibility to insecticides in synanthropic cockroaches: topical coating of prothorax of each insect with insecticides; contacting of insects with surfaces treated by insecticide solutions, emulsions, suspensions; dipping of insects into emulsions of different concentration insecticides, as well as spraying of insects with solutions and feeding with treated food.

At present there are biochemical tests for defining some resistant mechanisms. However, they are considered as amendment to traditional methods. Positive feature of biochemical tests is that insect samples can be kept even at extremely low temperatures and processed in remote laboratories. Main disadvantage of these tests is that at present they are not unified, their cost is relatively high, and highly qualified personnel is required to work with them.

The most efficient strategy of present-day disinsection is making integrated programs combining application of different agents: mechanical, biological, chemical means, genetic

* Tolerance *(tolerantia)* - the capacity of an organism to endure unfavorable environmental factors.

methods. Making of any programs is to be based on preventive measures which can help (without insecticides) to sanitate environment and to limit (or eliminate) reproduction and propagation of the arthropoda.

Resistance appears faster in restricted areas where there is no arthropoda influx from untreated territories. Resistance appears slower and its level is relatively low, when there is interchange between species from treated and untreated areas. Excess of recommended dosage of insecticides results in fast selection of resistant individuals.

Method of "multiple attack" is proposed against resistance and for reducing it, when rotation is used, i.e. alternation of insecticides with different mechanisms of action. In its turn it can result in fast formation of cross-resistance.

It is impossible to make unique sequence in applying pesticides for each case. Selection of strategies, insecticides is stipulated by local conditions; biology, ecology and behavior of dominating arthropoda species to be controlled; characteristic of objects to be treated; possibility to make sanitation, etc. For example, alternation of permethrin and temephos (diphos) destroyed temephos-resistant turkey gnats – onchocercosis transmitter in one of Africa regions.

Changing of methods for applying insecticides can affect more sensitive stages of insect development. For example, insecticidal baits and larvicides are recommended to be widely used to control resistant populations of house fly imagoes.

Using effectively and instantly acting sprayers allows resistant population to be quickly destroyed, however, in some cases application of insecticidal sprayers can forward fast selection of resistant individuals.

The scheme of disinsection measures to control cockroaches in large object can be as follows: 2-3 treatments are made with pyrethroids, the next two treatments are made by organophosphates, for further treatments digestive poisons only (baits with hydromethylnon, sulfluoramide and others) in combination with preparations based on boric acid and glue baits are used, then pyrethroid treatment can be made again.

Population recruitment after being reduced is very important. It should be taken into account that any attempt to destroy population of animals (arthropoda) is involving adaptive compensatory mechanisms, and homeostasis of destroyed system is restored.

Even the most effective methods and means of disinsection used for treating large objects (areas) cannot guarantee 100% destruction of arthropoda population, some individuals survive due to different biological reasons (individual poison-resistance, reduced food reflexes, excessive irritability, etc.). Numerous scientific and practical data testify that this remaining part of population can shortly restore its initial population. For example, if 5% of German cockroach population survives after disinsection in the object, then 4 months later (in absence of repeated treatments) the population will 1.2 times exceed the initial one, if 10% survives (90% effectiveness), 4 months later the original population will be doubled.

When the object is planned to be disinsected one should determine which preparations and methods of insect control were applied here and sensibility level of insects to the insecticide (or insecticides) meant to be used.

Possibility of irritability phenomenon to any preparation is to be considered in insects while applying insecticides. This reaction is known in malaria mosquitoes which leave treated surfaces before reaching lethal dose. This phenomenon is named as "stimulating-repellant action of insecticides". Irritability reaction is revealed in synanthropic cockroaches who avoid contacting surfaces treated with preparations based on cyhalothrin, deltamethrin, cypermethrin.

As toxic and environmentally hazardous insecticides are applied for disinsection, their selection and sequence of application require through knowledge in complex maximum control of the arthropoda in addition to early detection and exact resistance monitoring.

Chapter 22

22. Pesticide Safety Measures

Annually 500,000 cases of pesticide poisoning are registered in the world. Number of lethal cases is estimated as 5000 per year. Poisoning takes place under different contacts with pesticides. Workers come into contact when manufacture them and making preparations from them. Another possibility appears at pesticide application – both at loading of preparations and at direct applying to plants, the same can be said of air spraying. Besides, in some cases farm workers are to weed or gather crops after pesticide treatment of fields. In this case unprotected parts of the body are touched by pesticides. Large doses of pesticides can get into the human organism not only at direct contact with them, but while drinking contaminated water, or eating poisonous food, inhaling contaminated air, contacting objects treated with pesticides. 50% of foodstuffs sold in the USA contain pesticides in dangerous concentrations [189]. It is found that accumulation of some pesticides in grain can be the reason for their occurrence in processed products, as they accumulate not only in glumes but in endosperm as well. Therefore, maximum acceptable pesticide concentration in foodstuffs is fixed in majority of the world countries, their number exceeding no 0.01-5.0 mg per 1 kg of the product depending on toxicity and rate of pesticide decay.

To prevent workers from pesticide poisoning rates for applying preparations and their spray materials, number of treatments, “waiting” periods, i.e. time between the last treatment and harvesting, and time for working in pesticide-treated areas are to be observed. Public safety near treated areas is also to be secured. Sanitary gaps between treated areas and residential zones, recreational facilities, summer convalescence homes, and water supply sources are to be provided.

Places for storing preparations to be applied in private plots (for controlling mouse-like rodents inclusively) must not be inaccessible for children and domestic animals. Containers used for food and drinking must not be utilized for preparation keeping and applying.

Under centralized storage it is strongly prohibited to store pesticides in bulk in premises not intended for keeping them, in the open air, and to use for this purpose storehouses intended for keeping forage, foodstuffs, different household items and houseware.

All preparations must be received by storehouses in containers meeting normative and technical documentation. If their integrity is damaged pesticides are to be shortly placed into special package. It is strongly prohibited to leave pesticides spilled.

Empty returnable containers are to be returned to the supplier, disposable paper or wooden containers are to be burnt in special installations or in special cites. Using these containers for keeping water, foodstuffs, forage and for other household purposes is strongly prohibited. Ironically, these containers (metal or made from polymers) can be seen in unlikely places.

After finishing all works with pesticides it is recommended to take off working cloths, wash hands and face with soap, to rinse mouth and to take a shower (if possible).

The way of protection depends on possible input of contaminants into the organism (through respiratory organs or skin). Respirators should be used against dust or gases. Goggles are used to protect eyes. Workers should wear protective gloves, as well as protective clothing like overalls, coveralls and jackets. Leather or rubber footwear is to be used for legs protection.

Professional training of people directly participating in transportation and application of pesticides provides for detailed knowledge of used preparations and safety measures to handle them, including first-aid predoctor care in cases of poisoning.

Basic legislative documents are federal laws: "On safe handling of pesticides and agrochemicals", "On sanitary-and-epidemiologic welfare of the population", as well as sanitary rules and norms "Hygienic requirements to storing, application and transportation of pesticides and agrochemicals".

Sanitary regulations are to be observed in all industrial facilities, for all processes and equipment for storing, transportation and application of pesticides and agrochemicals by all individuals and legal entities applying these means, as well as planning, constructing and running industrial facilities, transportation means, technological, sanitary appliances for handling pesticides and agrochemicals.

Table 13 presents xenobiotic classification by toxicity level.

Table 13. Xenobiotic* classification by toxicity level

Toxicity level	Enteral introduction	Inhalation introduction	
	LD_{50} (*mg/kg*)	LC_{50} (*mg/l*)	MPC (*mg/m*^{3})
Extremely toxic	less than 15	less than 1	less than 1
Highly toxic	15 - 150	1 - 10	10
Moderately toxic	151 - 1500	11 - 40	100
Low toxic	over 1500	over 40	over 100

Special section of the Regulations is devoted to environmental management and output of quality foodstuffs indicating additional measures providing for maximum safe handling of pesticides and exclusion of accumulation of pesticides and other toxic elements in production of plant cultivation and cattle breeding.

Classification of chemical substances by hazard level is given in Table 14.

* Xenobiotic – foreign (not participating in plastic or energy metabolism) substance occurring in internal environment of the organism.

Table 14. Classification of chemical substances by hazard level (GOST 12.1.007-76)

Indicator	Class of hazard			
	1	2	3	4
MPC, *mg/m*^{3}	<0,1	0,1-1,0	1,1-10,0	>10,0
LD_{50} p/o* (*mg/kg*)	<15	15-150	151-5000	<5000
LD_{50} p/cut (*mg/kg*)	<100	100-500	501-2500	<2500
LC_{50}/2 h (*mg/m*^{3})	<500	500-5000	5001-50000	<50000
IPIT*	>300	300-30	29-3	>3

General pesticide warehouse rules:

- Unauthorized persons are allowed to enter the warehouse after preliminary registration in the office only.
- Smoking and working with open flame is prohibited.
- Special operations, e.g. working with open flame (welding and gas cutting), as well as other sparking operations (slotting, grinding and drilling) are allowed to be made from written permission of the authorized persons only.
- Warehouse employees are to keep working places clear and in good order and to inform senior authorities of possible accidents.
- Avoid direct contact with open preparations while working. Do not inhale dust, vapor and wear protective clothes and goggles.
- Food intake and smoking are allowed in special premises.
- Thoroughly wash your hands before food intake.
- After the work it is recommended to wash face thoroughly or take a shower.
- Carefully handle warehouse materials avoiding package damaging, or preparation spilling.
- The preparations are to be kept in specially marked places or storage platforms. Stacking is to be observed. The stored preparations are to be clearly seen.
- Escape routes, emergency exits and access roads for fire brigade are to be kept clear.
- In-house transport rules are to be observed.
- Inform immediately of damaged packages, preparation spillage and promptly eliminate them.
- Wastes of preparations, e.g. long-term or unused preparations and dirty packing or liquid-absorbing materials are to be destroyed in accordance with special instruction.
- Local conditions should be considered while observing application instructions and safety rules.

---

* p/o – per os.

* IPIT - index of potential inhalation toxicity. Defined as ratio of maximum possible concentration of toxicant vapor in the air to average lethal concentration.

Chapter 23

23. Control for Content of Pesticide and Insect-Pests in Grain and Cereals

23.1. Removal of Pesticide Residues from Grain and Cereals

Grain treated with pesticides can be used as food or fodder or to be processed with further mandatory control for pesticide residues in accordance with official methods for determination. To decrease pesticide residues in grain and cereals to sanitary-permissible standards they are to be passively degassed after fumigation.

Degassing rate depends on conditions (temperature, air and product moisture) of the crop to be disinfected.

In urgent cases and for fast realization active aeration can be made through stationary or mobile ventilation installations and fans.

Active degassing cannot be applied, as it may result in grain cooling. When outdoor temperature is below the temperature of grain, and in all cases when fumigation takes place at outdoor temperature of 12°C, the grain is degassed while heating the air to be input to the bulk grain. Non-degassed grain and cereals are prohibited to be cooled.

Grain, flour and groats treated with methyl bromide are degassed at least five days in warehouses and grain elevators under passive aeration; 2 hours in storage silos and gas chambers equipped with recirculation plants under ventilation; and 10-15 hours in storage silos to be ventilated.

After expiration of these terms degassing results are checked by indicated burners in warehouse or storage silo places free from grain (production). When burner indicators are negative air samples are taken from bulk grain (intergrain air spaces) or from stacks with cereals for quantification of methyl bromide in the air.

Working in granaries, grain-processing plants and warehouses is allowed when methyl bromide content in air samples exceeds no 0.05 mg/l. Air samples in empty premises, bulk grain (intergrain air spaces) and in cereals for completeness of degassing are taken in the following way.

Air samples are taken into 0.5 l gas pipettes from poorly ventilated places of the premises, where degassing conditions by haloid burners were the worst.

Grain, flour, groats and mixed feed treated with methyl bromide are allowed to be used as food or fodder or to be processed with further mandatory control for inorganic bromide content.

When applying preparation 242 ending of degassing of bread grain and food legumes to be processed, as well as of legume seeds is defined organoleptically by absence of fumigant smell on the whole and in the ground grain.

Organoleptic method allows preparation 242 to be revealed in fumigated grain at its content of up to 1 g per 1 t under permissible dose of 2 g per 1 t. Feed grain and bread grain to be realized to the consumers, food legumes realized without processing, as well as cereals are to be chemically controlled for preparation 242 residues. They are allowed to be realized only when no fumigant residues are present.

Realization of seed grain disinfected with methallyl chloride and dichloroethane is allowed in absence of fumigant smell on the whole and in the ground grain.

Degassing level of feed grain and bread grain disinfected with methallyl chloride, as well as of seed grain if it was not used as intended and was used as food or fodder, and disinfected with dichloroethane is defined chemically.

Sampling of grain and cereals for completeness of degassing is made in accordance with the standard. Organoleptical analysis of grain is made promptly after testing.

Initial samples characterizing upper, middle and lower layers of the bulk are formed from tested ones in accordance with the standard, from which by two samples weighting 300 g each are taken for chemical analysis.

In grain elevators samples are taken in each bin separately making initial samples per each 3,000 tones of grain of similar crop. If elevator contains grains of different crops weighing over 3,000 tones, each of them is twice sampled for analysis.

To analyze cereals from samples characterizing separate stack two samples weighting 150 g each are taken.

For prompt realization of grain treated with organophosphate pesticides grain is transporting by elevator or warehouse technological lines (including grain separators).

23.2. Carbophos Quantification in Grain

This method is specific and based on extraction of carbophos from grain by carbon tetrachloride with further decomposition of extracted carbophos by alkali liquor over ethanol, thus forming sodium dimethyldithiophosphate and sodium fumarate. Optical density of this solution is defined by photoelectric colorimeter (PhEC). Detection sensitivity is – 0.02 mg per sample. Colored plant pigments and admixtures, which can be formed under alkaline hydrolysis of the carbophos and prevent from further determination, are removed after acidation with hydrochloric acid by carbon tetrachloride extraction. Adding of cupric sulphate to sodium solution of dimethyldithiophosphoric acid forms a complex, which solution in carbon tetrachloride is painted yellow. Carbophos weighted for analysis containing exactly 50 g of active agent is placed into prior weighted weighing bottle of the analytical balance. For example, if carbophos contains 50% of the active agent, then weight of quantity for analysis will be $x = (50 \cdot 100):50 = 100.0$ mg. Weighted amount is quantitatively taken to 50 ml volumetric flask. Over 2 ml of ethanol is placed into the weighing bottle, the preparation is dissolved and the solution is poured into the flask. This operation is repeated about 10 times. Then the weighing bottle is washed by ethanol from outside. After that the flask is filled up with ethanol to the mark. The obtained solution No.1 is thoroughly mixed (~1 ml of the solution contains 1 mg of the carbophose active agent). Standard solution No.2 is prepared (1 ml contains 0.1 mg of the carbophose active agent). By 100 ml of carbon tetrachloride is poured into separating funnels, and each of them except for the first one is added by solution No.2 (2.5; 5, 10, 15, 20 and 25 ml). Each funnel is added with 25 ml of ethanol and solutions are shaken. Further operation is made with up to one-two funnels simultaneously. The funnels are additionally filled with 1 ml of 6 N caustic liquor, and shaken within 1 minute. Then the funnels are shortly added with 75 ml of sodium chloride brine prior cooled to at least 15°C and are thoroughly shaken within 1 minute. The funnels are placed into the rack for full phase separation. Then lower layer of the liquid containing carbon tetrachloride is isolated. The remaining upper layer is added with 25 ml of carbon tetrachloride and 1 ml of hydrochloric acid. The funnel is thoroughly shaken within 1 minute. After phase separation the lower layer is isolated as full as possible. If at layer boundary there are emulsion bubbles or carbon tetrachloride is colored yellow, phase are separated additionally with one carbon tetrachloride (adding no hydrochloric acid). Emulsion bubbles are taken away together with the lower layer and isolated. Exactly 15 ml of carbon tetrachloride and 2 ml of copper sulfate solution are added to the funnels. The funnel is shaken within 1 minute and the layers are left for separation. The lower layer is poured into 50 ml conical flask. The upper layer is isolated. The lower layer is added by 2-3 g of anhydrous sulfate of sodium, filtered and colorimetered by PhEC. Each solution is colorimetered at least three times and all spots are marked on calibrating diagram. Weighted ground grain of 100 g is placed into 500 ml conical flask, 100 ml of carbon tetrachloride is added, and plugged, it is strongly shaken for 5 minutes and left for 24 hours at room temperature. Then the flask is shaken, the solution is paper-filtered into 250 ml flask. Volume of the extract is measured, and diluted by carbon tetrachloride to get 100 ml. The resultant solution is taken to the separating funnel. 25 ml of ethanol is added and the solution is shaken. Further analysis is made like calibrating diagrams. PhEC results (number of points on the right drum) are plotted along ordinate axis of the calibrating diagram. From this point perpendicular is plotted until it intersects the diagram line. From the point of intersection the perpendicular is extended to x-coordinate. Amount of active substance of carbophos is calculated in the point of intersection. Average value is calculated from three measurements. Carbophos content (mg/kg) in grain is calculated by the formula:

$$x = \frac{100b}{0{,}1a} = 1000\frac{b}{a},$$

where, 100 – amount of carbon tetrachloride taken for extraction, ml; a – amount of extract from grain after filtration, ml; 0,1 – amount of grain taken for analysis, kg; b – amount of carbophos active agent calculated by calibrating diagram, mg.

23.3. Colorimetric Estimation of Organophosphorous Pesticides (Carbophos, Metathion, Volaton and Dichlorvos) in Grain and Flour by Total Phosphorus

This method is based on extraction of pesticide from grain and flour by carbon tetrachloride, its oxidation till *ortho*-phosphate by ammonium persulphate with further colorimetric estimation by blue phosphoro-molibdenic heterocomplex. Optical density of the solution is measured by photoelectric colorimeter PhEC-M with red (No.8) color filter.

0.2742 g of monosubstituted potassium phosphate is weighted in preliminary weighting bottle of the analytical balance, quantitatively taken to 250 ml volumetric flask, dissolved in the distilled water acidated by several drops of sulfuric acid, and volume of liquid in the flask is added till the mark on the flask and thoroughly mixed (standard solution No.1).

The solution is effective within two-three weeks. 10 ml of standard solution No.1 is taken by pipette and introduced into another 250 ml volumetric flask, added by the distilled water till the mark and mixed (standard solution No.2).

All test tubes except for the first one are added with 0.005; 0.1; 0.2 and 1.0 ml of standard solution No.2, corresponding to 05; 1.0; 2.0 and 10.0 microgram of phosphorus.

Test tubes are added by 2 ml of 0.25 M solution of ammonium persulphate, placed on metal support into the water bath with boiling water and heated within ~10 minutes. Then it is promptly cooled in water with ice or in cold water bath and added with 3 ml of 0.2 M of urea solution and thoroughly mixed. Test tubes are again placed into the water bath for 5 minutes and quickly cooled, and are added by 1 ml of 2,5%-solution of ammonium molybdate and by 1 ml of 2% of ascorbic acid, mixed and placed into the boiling water bath for 2 minutes, and cooled again. The solution in test tubes is colored dark blue. Test tubes are added by 3ml of isobutyl alcohol and shaken.

Coloring of the solution passes to the upper layer (isobutyl alcohol), whose optical density is measured by PhEC with filter No.8 (597 nm) in cell of 5 mm thickness. Solutions (at least two) which were not added by standard solution No.2 are used as control ones. Under the received data calibrating diagram is plotted.

Weighted amount of 50 g of grain (flour) is placed into 300-500 ml conical flasks and is taken from each analyzed sample in the counter balance. Simultaneously, similar weighted amount of grain (flour) is placed into two flasks but containing no pesticide. Each flask is added by 30 ml (by 75 ml when analyzing flour) of carbon tetrachloride, shaken manually or on shaker and extracted within 20 minutes. Carbon tetrachloride is filtered out. By 20 ml of the solution is taken from filtrates into 50-100 ml flasks and placed into the water bath. Carbon tetrachloride is evaporated till 1-2 ml volume.

The solution is quantitatively taken to test tubes. The flasks are flushed out with carbon tetrachloride 3-5 times, for the general volume of the solution in the test tube to be 5 ml. Test tube in metal support with the pesticide solution in carbon tetrachloride is placed into the water bath (70°C) with its further heating until it boils. Carbon tetrachloride is evaporated from the test tubes until dry. Then evaluation is made as described for calibrating diagram.

Content of pesticide (mg/g) in grain (flour) is calculated by the formula:

$$x = \frac{abK}{cgK_1},$$

where a – amount of preparation found by the calibrating diagram, microgram; b – amount of carbon tetrachloride taken for extraction of preparation from grain or flour, ml; c – amount of extract taken for analysis, ml; g – weight of weighted product, g; K – phosphorus / pesticide translation factor; K_1 – preparation extraction factor for grain or flour.

Factors for translation of phosphorus to different pesticides, as well as extraction factors for different organophosphorous preparations from grain and flour are given in Table 15.

Table 15. Factors for translation

Pesticide	Phosphorus / pesticide translation factor	Preparation extraction factor	
		from grain	from flour
Carbophos	10.66	-	0.95
Dichlorvos	7.13	0.54	0.80
Metathion (methylnitrophos)	8.95	0.56	0.98
Volaton (phoxim)	9.17	0.58	-

23.4. Estimation of Grain Infestation Affected by Grain Products Pests

Grain infestation is defined by number of pests per 1 kg of grain. Estimation is made over grain to be delivered to warehouse and in stored grain when necessary. Grain is sampled when bulk grain exceeds 1.5 m from each section in 100 m^3 of three layers: upper – 10 cm from the surface, central – from the middle of the bulk, and lower – at the bottom; when bulk height is less than 1.5 m – from upper and lower layers. Each layer is sampled by 1 kg separately. Infestation of each grain lot is estimated by the most infested sample.

First, mite infestation is estimated. For this purpose 1 kg of grain sample is screened manually through 1.5 mm meshes within 2 minutes (at 120 circular motions per minute). If sidewall diameter of the sieve is 20 cm, the grain is screened in three lots (about 1/3 of the sample per lot). When sidewall diameter is 30 cm all grain is screened simultaneously.

If grain temperature is below 5°C, the sieve is warmed at 25-30°C within 10-20 minutes, for all insect to start moving.

All sieved grain is scattered in thin layer over the glass with black paper laying under it, and the grain is examined through 5-10-fold magnifying conical lens. Number of mites is counted and infestation level of this lot is estimated by Table 16.

Table 16. Grain infestation level

Grain infestation	Number of mites per 1 kg of grain
I	1-20
II	Over 20
III	Mites make continuous tomentose layer

To estimate grain infestation by other insect-pests 1 kg of grain is screened manually through the sieve with 1.5-2.5 mm meshes. All grain is sorted out isolating pests (dead ones are considered as black dockage and ignored at grain infestation). There are three levels of infestation with granary and maize weevils (see Table 17).*

Table 17. Level of grain infestation

Infestation level	Number of pests per 1 kg of grain
I	1-5
II	6-10
III	Over 10

Latent form of infestation with weevils is defined by method of Alpha-Beta search: weighted quantity of 15 g is taken from grain sample, with all dockage and weevil-damaged grains being taken away, and the grain sample is placed onto the small net in tin case, plunged into the cup of water of 30°C for 1 minute and taken for 20-30 sec (wheat and rye for 45 sec) into 1% solution of potassium permanganate.

Plugs which are used by insects to paste their eggs are painted black therewith, and undamaged surface of grain retains its usual color. Potassium permanganate is washed from grain by plunging nets with grain into the cold water or into sulfuric acid solution with hydrogen peroxide (1 cm^3 of 3% hydrogen peroxide solution is added per each 100 cm^3 of 1% sulfuric acid solution). After 20-30 sec of exposure in this solution sound grain becomes clear and infested grain contains convex black plugs. When corn is analyzed for infestation each tenth ear of the combined sample is carefully examined under 4.0-4.5 magnifying lens and number of living pests is counted.

To define mite infestation ten ears in pairs are taken from the combined sample and slightly knock one on another over the black paper. Residues left on the paper are examined through the lens, isolating and counting mites. Degree of infestation, pest species and their number are indicated in the reporting documents [190].

Under the layer sampling of grain according to GOST 13586.3-83 the analysis is made by average sampling taken separately from each layer and infestation is defined by sample with the maximum number of pests. Balls of grain covered with butterfly caterpillars are sorted out manually. Revealed pests are added to total pests discovered in average sampling. After sorting out the balls the average sample of grain is weighted and then screened through

* If grain is infested with several pest species it is screened simultaneously through two sieves with 2.5 and 1.5 mm meshes.

sieving stack with 1.5-2.5 mm diameter of meshes manually within approx.2 minutes at 120 motions per minute or mechanically in accordance with the attached manual. If grain temperature is below 5°C, the received grain and sieve residues are warmed at 25-30°C within 10-20 minutes for activation of insects. 2.5 mm sieve residues are placed onto the table for analyzing, leveled in thin layer and sorted out by spatula revealing large insects: wheat beetle, yellow mealworm and lesser mealworm beetles, white-marked spider beetle and others. 2.5 mm sieve residues are placed on white glass of the table for analyzing, and 1.5 mm sieve residues are placed on black glass scattering by rarified thin layer; 1.5 mm sieve residues are studied under the magnifying lens. Small pests are revealed thereby: granary and maize weevils, lesser grain borer, red flour and confused flour beetles, saw-toothed grain and rust red grain beetle, flour and elongated mites, and others. Dead pests as well as living field pests not damaging stored grain are considered as black dockage and ignored at grain infestation. Number of living pests received is counted per 1 kg of grain.

Grain infestation in the latent form is determined by splitting grains or by method for "plugs" coloring (closed holes after oviposition). Infestation by splitting is defined by weighted portion of 50 g taken from the average sample. 50 whole grains are randomly taken from the weighted portion and split with a scalpel tip along the furrow. Split grains are examined under the lens and living insects are counted at different stages of development. Infestation by "plugs" coloring is defined by weighted portion of 50 g taken from the average sample. 250 whole grains are randomly taken from the weighted portion and plunged into a cup of water at approx. 30°C for 1 minute. The grain starts to swell and the "plugs" size is increased simultaneously. Then net of grains is placed for 20-30 sec into 1% fresh solution of potassium permanganate (10 g of $KMnO_4$ per 1 liter of water). Not only the "plugs", but the damaged grain surfaces are dark painted. Excess of paint is removed from grain surface by plunging the net of grain into cold water for 20-30 sec, therewith sound grains restore their natural color and the infested ones retain dark convex "plug". Grains taken from water are promptly examined on the filter paper. Infested grains are to be counted immediately without leaving the grains drying out, otherwise coloring of the "plugs" disappears. Infested grains are characterized by round convex spots of about 0.5 mm, by the "plugs" evenly dark colored left by the weevil female after eggs laying.

Grains considered as non-infested:

- having round spots with intensively colored edges and light center which are places of weevil feeding;
- having spots of irregular shape in places of mechanical damage of grain. Infested grains are cut and number of living larvae, pupae or weevil beetles is counted.

Content of grains infested latently (X_3) in per cent is calculated by the formula:

$$X_3 = \frac{n_3}{n} \bullet 100,$$

where n_3, - a number of infested grains, pieces; n – a number of grains selected for analysis, pieces.

Samples are to be protected against extreme temperature and moisture as well as against direct sunlight to minimize changes in the insect population due to their death from heating or dehydration. If there many insects and they are active, redresser sieve and the pallet are used to isolate insects from the sample. Grain layer in this case must not exceed three grains, otherwise the sample is divided. After screening when there are not many insects and they are less active, the grain is placed in one layer on a tray or in a cup, and all revealed insects are picked out with entomological pincers or brush. Species of insects found in the sample are determined and number of each insect species (of larvae and pupae, if possible) is registered. If necessary, number of dead and living insects is indicated separately. After insect determination the sample is returned back to the same container to be sealed and placed to the cell or incubator. If grain moisture exceeds 15%, relative humidity in the cell or incubator where the sample is placed is to be maintained at 60-65%. Incubation continues for 36 days. Actual length of incubation period depends on temperature at which the sample is kept, kind of used grain and species of insects present in grain. Recommended length of the incubation period for some insect species is given in Table 18.

Table 18. Incubation periods in days to reveal hidden stages of insect development in cereals and legumes samples, depending on temperature

Species	Russian name	Incubation period in days at	
		25°C	30°C
Acanthosceldes obtectns	Фасолевая зерновка	56	42
Aracocerus fasciculatus Deg	Какаовый ложнослоник	84	56
Callosobruchus maculates (F)	Четырехпятнистая зерновка	49	35
Rhyzopertha dominika (F)	Зерновой точильщик	70	49
Sitophilus granaries (Z)	Амбарный долгоносик	56	42
Sitophilus oryzae	Рисовый долгоносик	56	42
Sitophilus zeamais	Кукурузный долгоносик	56	42
Sitophilus cerealella (Oliv)	Зерновая моль	56	42
Zabroter subfasciatus (Boh)	Бразильская зерновка	56	42

If the sample contains more than one insect species, the longest incubation period is chosen.

Number of insects initially revealed in each weighted portion is registered by species and stages of development (i.e. adult species, pupae, larvae and eggs), if necessary number of dead and living insects is indicated. Total number of insects in weighted portions is counted and, initially revealed infestation is expressed as number of insects per 1 kg of grain separately by each species and stage of development. Number of insects revealed in each weighted portions under further examination is registered by each species and stage of development summarizing results for all weighted portions. After the last examination the results of all analyses are summarized and, hidden infestation by number of insects per 1 kg of grain is expressed separately by species and stage of development considering laboratory samples. If, within first 7 days after the analysis there appear adult species in the weighted portion, then before determining hidden infestation the number of insect species appeared after the period indicated in the Table is subtracted from the total number of the same insects. In this case, insects appeared after the incubation period are supposed to be the offspring of adults constituted the initially revealed infestation stipulating the total infestation during sampling.

The results of counting the appearing individuals permit to consider age distribution of each insect species at the time of sampling. The data plotted on the diagram show changes of ratios of different stages of insect development (from eggs till adults) in equal intervals. High content of young individuals (insects appeared late) indicates of the population increase in the sampling zone, and lower content – of the population decrease. Number of found insects depends on temperature for storing the product. At 15°C and below none of insect species mentioned in the Table cannot be reproduced quickly. At 25°C and higher presence of just one species from the above Table in 1 kg of grain is a serious hazard.

23.5. Carbon Dioxide Emission Method

This method is based upon measuring amount of carbon dioxide liberated for a certain period in the process of metabolic reaction occurring in insect organism. Rate of exchange reactions in dry grain or cereals is very low. Insect metabolism takes place faster enough, so carbon dioxide emission in dry grain or cereals can be considered as infestation feature, provided grain infestation with this gas was prevented and it was not absorbed by grain surface.

Amount of carbon dioxide emission is determined by gasometrically or by using devices for registration of infrared spectra.

23.6. Ninhydrin Method

Ground powder of the analyzed grain is placed on ninhydrin-soaked paper. After dry grain grinding amino acids contained within insects react with ninhydrin and make red spots on the paper, while amino acids of grain do not react.

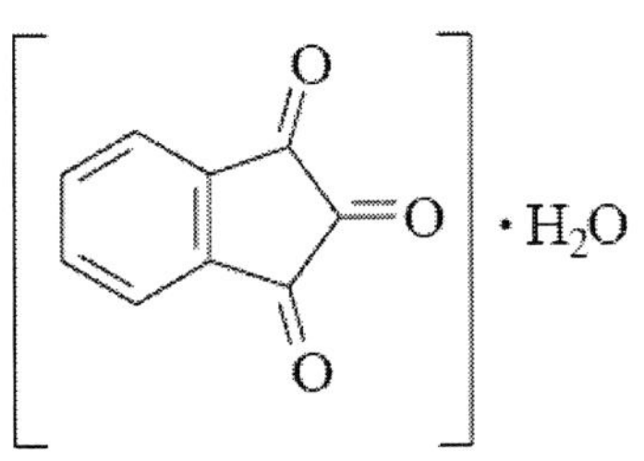

Ninhydrin

High humid grain can react itself in 2-3 days, so when making analysis using this method grain moisture is to be strictly controlled. Number of spots indicates hidden infestation of the sample. It should be taken into account at analyzing that amino acids contained in the skin of fingers also react with ninhydrin and make the skin red (it can be avoided by putting on surgical gloves or using pincers). At 20°C and higher red spots appear in one hour, and reach maximum brightness in 24 hours. Al lower temperatures or when it is necessary to get coloring faster, the paper can be heated in the oven at 50°C, or over the spirit lamp, or electric lamp. Appeared red spots are marked by circling with the pencil. One should be very attentive as the spots can be located close to each other and almost coincide. Spots of other colors are not considered. Eggs and small larvae cannot always be revealed by this method. Therefore, the grain actually free from insects is recommended to be repeatedly analyzed in 2-4 weeks, for the insect development occurred. Efficiency of this method greatly depends on insect species and size, and on kind of grain to be analyzed. Positive results indicate that the grain storage is unsafe. One red spot corresponds to one insect in the weighted portion, and relatively small number of spots randomly scattered on the paper (if there are similar results in other weighted portions) indicates light or average infestation level. Large number of red spots indicates strong infestation and the necessity to be treated promptly. However, before taking any measures one should determine whether the grain was earlier effectively treated and if so, when it was treated. It is due to the fact that dead insects continue to give positive reaction until liquid in their bodies dries out, which can last for several weeks in large insects.

23.7. Flotation Method

Immersion of the mixture of sound and infested grain into the water solution of sodium silicate, ammonium nitrate or glycerol makes it possible to separate the sound grain, as it sinks in the solution with certain density. The infested one remains on the water surface as hidden infestation decreases its weight. The division in indistinct as the grain with larvae tends to go dawn at early stages of development, and infested one with cavities inside can go up.

23.8. X-Ray Method

Exposition is made by soft X-radiation and then the film is examined visually for insects in grain. To get roentgenogram the weighted grain is leveled by one grain between X-ray source and X-ray film. Eggs and small larvae can be detected just occasionally when examining general result of exposition. X-ray method fails to reveal each egg and larva at the early stage of development.

References

[1] Cherkovkaya A.Ya., Zeltova S.A., Nemchinova M.M. Reference-book for protection of grain and products made of grain. M.: Kolos. 1984, P. 11-19.

[2] Mordkovich Ya.B., Zaets V.G., Risha A., Emmanuel E.I. // *Herald of theRussian Academy of Agricultural Sciences.* 1997. № 1, C.61.

[3] Mohammad I., Ahmad G., Javed I. // *Pakistan J. Zool.* 1992. V.24. N 1. P.79. [Abstract Journal Biology. 95.10-04И3.381].

[4] Zhantiev R.D. Dermestid beetles of the USSR fauna. M.: MGU. 1976. P.182.

[5] Levinson H.Z., Levinson A.R., Jen T.-L., Williams J.L.D., Kahn G., Francke W. // *Naturwissenschaften.* 1978. V.65. P.543.

[6] Odinokov V.N., Burov V.N., Kukovinets O.S., Ishmuratov G.Yu., Shamshev I.V., Selitskaya O.G., Zainullin R.A. Semiochemicals for protection of grain and grain products. Ufa: Gilem. 2005. P.232.

[7] Cross J.H., Byler R.C., Cassidy R.F., Jr. Silverstein R.M., Greenblatt R.E., Burkholder W.E., Levinson A.R., Levinson H.Z. // *J. Chem. Ecol.* 1976. V. 2. P.57.

[8] Yarger R.G., Silverstein R.M., Burkholder W.E. // *J. Chem. Ecol.* 1975. V. 1. P.323.

[9] Nara J.M., Lindsay R.C., Burkholder W.E. // *J. Agric. Food Chem.* 1981. V. 29. P. 68.

[10] Mori K. // *Tetrahedron.* 1974. V. 30. P. 3817.

[11] Mori K., Kuwahara S., Levinson H.Z., Levinson A.R. // *Tetrahedron.* 1982. V. 38. P. 2291.

[12] Silverstein, R.M., Rodin, J.O., Burkholder, W.E., and Gorman, J.E. Sex attractant of the black carpet beetle.// Science. 1967. V.157. P.85-87.

[13] Kodama H., Ono M., Kohno M., Ohnishi A. // *J. Chem. Ecol.* 1987. V. 13. P. 1871.

[14] Kodama H., Mochizuki K., Kohno M., Ohnishi A., Kuwahara Y. // *J. Chem. Ecol.* 1987. V. 13. P.1859.

[15] Kuwahara Y., Fukami H., Howard R., Ishii S., Matsumura F., Burkholder W.E. // *Tetrahedron.* 1978. V. 34. P.1769.

[16] Hodges R.J., Hall D.R., Mbugua J.N., Likhayo P.W. // *Bull. Entomol. Res.* 1998. V. 88. P. 131.

[17] Cork A., Hall D.R., Hodges R.J., Pickett J.A. // *J. Chem. Ecol.* 1991. V. 17. P. 789.

[18] Bashir T., Birkinshaw L.A., Farman D., Hall D.R., Hodges R.J. // *J. Stored Prod. Res.* 2001. V. 39. P. 159.

[19] Williams H.J., Silverstein R.M., Burkholder W.E., Khorramshahi A. // *J. Chem. Ecol.* 1981. V. 7. P. 759.

[20] Cogburn R.R., Burkholder W.E., Willams H.J. // *Environ. Entomol.* 1984. V. 13. P. 162.

[21] Pierce A.M., Pierce H.D., Jr. Borden J.H., Oehlschlager A.C. // *J. Chem. Ecol.* 1991.V. 17. P. 581.

[22] Oehlschlager A.C., King G.G.S., Pierce H.D., Jr. Pierce, A.M., Slessor K.N., Millar J.G., Borden J.H. // *J. Chem. Ecol.* 1987. V. 13. P.1543.

[23] Pierce A.M., Pierce H.D., Jr., Borden J.H., Oehlschlager A.C. // *J. Chem. Ecol.* 1991. V. 17 P. 581.

[24] Oehlschlager A.C., King G.G.S., Pierce H.D., Jr. Pierce A.M., Slessor K.N., Millar J.G., Borden J.H. // *J. Chem. Ecol.* 1987. V. 13. P. 1543.

[25] Pierce A.M., Pierce H.D., Jr. Oehlschlager A.C., Czyzewska E., Borden J.H. // *J. Chem. Ecol.* 1987. V. 13. P.1525.

[26] Pierce A.M., Pierce H.D., Jr. Oehlschlager A.C., Borden J.H. // *J. Agric. Food Chem.* 1985. V. 33. P. 848.

[27] Pierce A.M., Pierce H.D., Jr., Borden J.H., Oehlschlager A.C. // *J. Chem. Ecol.* 1991. V. 17 P. 581

[28] White P.R., Chambers J., Walter C.M., Wilkins J.P.G., Millar J.G. // *J. Chem. Ecol.* 1989. V. 15. P. 999.

[29] Millar J.G., Oehlschlager A.C. // *J. Org. Chem.* 1984. V. 49. № 13. P. 2332.

[30] Arnaud L., Lognay G., Verscheure M., Leenaers L., Gaspar C., Haubruge E. // *J. Chem. Ecol.* 2002. V. 28. P. 523.

[31] Phillips T.W., JiangX.-L., Burkholder W.E., Phillips J.K., Tran H.Q. // *J. Chem. Ecol.* 1993. V. 19. P. 723.

[32] Zakladnoi G.A., 1983. Protection of plants and products made of plants against pests. M.: Kolos. P.216.

[33] Endt D.W., Wheeler J.W. // *Science*. 1971. V. 172. P. 60.

[34] Engelhardt M., Rapoport H., Sokoloff A. // *Science*. 1965. V. 150. P. 632.

[35] Arnaud L., Lognay G., Verscheure M., Leenaers L., Gaspar C., Haubruge E. // *J. Chem. Ecol.* 2002. V. 28. P. 523.

[36] Arnaud L., Lognay G., Verscheure M., Leenaers L., Gaspar C., Haubruge E. // *J. Chem. Ecol.* 2002. V. 28. P. 523.

[37] Landor P.D., Landor S.R., Mukasa J. // *J. Chem. Soc. Chem. Comm.* 1971. P. 1638.

[38] Shu S., Koepnick W. L., Mbata G. N., Cork A., Ramaswamy S. B. // *J. Stored Prod. Res.* 1996. V. 32 P. 21.

[39] Phillips J.K., Chong J.M., Andersen J.F., Burkholder W.E. // *Entomol. Exp. APl.* 1989. V. 51. P. 149.

[40] Phillips J.K., Miller S.P.F., Andersen J.F., Fales H.M., Burkholder W.E. // *Tetrahedron Letters*. 1987. № 28. P. 6145.

[41] Vick K.W., Su H.C.P., Sower L.L., Mahany P.G., Drummond P.C. // *Experientia*. 1974. № 20. P. 17.

[42] Ohshima M., Kuwahara Y., Matsuyama S., Suzuki T. // *J. Pestic. Sci.* 1993. V.18. P.163.

[43] http://www.nysaes.cornell.edu/pheronet/phlist/tineola.html#li12

[44] Yamaoka R., Shiraishi Y., Ueno T., Kuwahara Y., Fukami H. // *Mass Spectro*. 1985. V. 33. P.189.

[45] Yamaoka R., Ueno T., Fukami H. // *Proc. 29th Conference Mass Spectro*. 1981. V. 7. P. 753.

[46] Turker L., Togan I., Ergezen S., Ozer M. // *Apidologie*. 1993. V. 24. P.425.

[47] Mori K., Harada H., Zagatti P., Cork A., Hall D.R. // *Liebig's Ann. Chem*. 1991. P. 259.

[48] Hall D.R., Cork A., Lester R., Nesbitt B.F., Zagatti P. // *J. Chem. Ecol*. 1987. V. 13 P. 1575.

[49] Read J.S. Haines C.P. // *J. Chem. Ecol*. 1979. V. 5. P. 251.

[50] Cox P. D. // *J. of Stored Products Research*. 1974. V. 10. № 1. P. 43.

[51] Kehat M., Eitam A., Blumberg D., Dunkelblum E., Anshelevich L., Gordon D. // *Phytoparasitica*. 1992. V. 20. P. 99.

[52] Brady, U.E., Daley, R.C. // *Ann. Entomol. Soc. Am*. 1972. V. 65. P. 1356.

[53] Tóth M., Repasi V., Szöcs G. // *Acta Phytopathol. Entomol*. Hung. 2002. V. 37. P. 375.

[54] Zhu J.-W., Ryne C., Unelius C.R., Valeur P.G., Löfstedt C. // *Entomol. Exp. APl*. 1999. V. 92. P. 137.

[55] Teal P.E.A., Heath R.R., Dueben B.D., Coffelt J.A., Vick K.W. // *J. Chem. Ecol*. 1995.V. 21. P. 787.

[56] Mudd A. // *J. Chem. Ecol*. 1985 V. 11. № 51. P. 57.

[57] Kuwahara Y., H. Hara S. Ishii, Fukami H. // *Agric. Biol. Chem*. 1971. V. 35. P. 447.

[58] Brady U.E. // *Life Sciences*. 1971.V. 10. P. 797.

[59] Zhu J.-W., Ryne C., Unelius C.R., Valeur P.G., Löfstedt C. // *Entomol. Exp. APl*. 1999. V. 92. P. 137.

[60] Patent 136495 CzSSR; Chem. Abstr. 1971. V. 75. 4029.

[61] Vondrachekova M., Sedmera P., Musilek V. // *Collect. Czech. Chem. Commun*. 1983. V. 48. P. 1508.

[62] Zakharychev V.V., Kovalenko L.V. // *Successes of Chemistry*. 1998. V. 67. No. 6. P. 595-605.

[63] Beautement K., Clough J.M., de Fraine P.J., Godfrey Ch.R.A. // *Pestic. Sci. 1991*. V. 31. P. 499.

[64] Glough J.M., de Fraine P.J., Fraser T.E.M., Godfrey Ch.R.A. // *Am. Chem. Soc. Symp. Ser*. 1992. V. 504. P. 372.

[65] Eikhler V. Poisons in our food. M.: Mir. 1993.

[66] Jean Dorst. Before the Nature will die, M.: Progress. 1968.

[67] Melnikov N.N. Pesticides. Chemistry, technology and application. M.: Chemistry. 1987. P. 400.

[68] Ho Tse-Lok, Din Zia-Ud. // *Syhth. Commun*. 1982. V. 12. No. 3. P. 257.

[69] Patent of India 154394 (1984).

[70] Kukovinets O.S., Zainullin R.A., Galin F.Z., Shereshovets V.V., Kashina Yu.A., Akhmetov A.M., Kunakova R.V., Tolstikov H.A. // *Journal of Organic Chemistry*. 2001. V. 37. Issue. 2. P. 251.

[71] Galin F.Z., Rakhimov R.G. Natyral pyrethrins and their analogs – highly effective means for plant protection. 2000. Ufa: Reactive. P.22.

[72] Shipov A.E. Development of industry exercisable mixtures based on domestic universal synergist (Sh-294) and pyrethroid insecticides to control household pests and agricultural insect-pests. Institute for hetero-organic compounds named after A.N. Nesmeyanov RAS.

[73] Zakharychev V.V., Kovalenko L.V. // *Successes of Chemistry*. 1998. V. 67. No. 6. P. 595.

[74] E. Geoffrey // *Ann. Inst. Colon*. 1895. V. 2. P. 1

[75] K. Nagai // *J. Tokyo Chem. Soc*. 1902. V. 23. P. 740.

[76] L. Crombie // *Nat. Prod. Rep*. 1984. V.1. P. 4.

[77] Belanger A., Berney D.J.F., Borschberg H.J., Brousseau R., Doutheau A., Durand R., Katayama H., Lapalme R., Leturc D.M., Liao C.C., Maclachlan F.N., Maffrand J.P., Marazza F., Martino R., Moreau C., St-Laurent L., Saintonge R., Soucy R., Ruest L. and Deslongchamps P. // *Can. J. Chem*. 1979. V. 57. P. 3348.

[78] Grapov A.F. // *Journal of D.I.Mendeleev All-Union Chemical Society*. 1984. V.29. P. 40.

[79] Mitsudera H., Konisi H. // *J. Pesticide Sci*. 1991. V. 16. P. 987.

[80] Kato A., Ishimaru M., Hashimoto Y., Mitsudera H. // *Tetrahedron Lett*. 1989. V. 30. P. 3671.

[81] Odinokov V.N., Kukovinets O.S., Zainullin R.A., Tolstikov H.A. // *Successes of Chemistry*. 1992. V. 61. No. 7. P. 1332.

[82] Bowers W.S., Ohta T., Clere J.S., Marsella P.A. // *Science*. 1975. V. 193. P. 542.

[83] Richard Dawkins. The Selfish Gene. M.: Mir. 1993.

[84] Chemistry and provison of mankid with food. Translated from English. M.: Mir. 1986. P.116.

[85] Nakatani N., Inatani R., Ohta H., Nichioka A. // *Environ Health Perspect*. 1986. V. 67. P. 135.

[86] Miyakado M., Nakayama I., Ohno N., Yoshioka H. // *Curr. Themes Trop. Sci*. 1983. V. 2. P. 369.

[87] Goodwin T.W., Horn D.H.S., Karlson P. // *Nature*. 1978. V. 272. P. 122.

[88] Barbier M. Introduction to chemical ecology. M.: Mir. 1978. P.232.

[89] Harborn D. Introduction to ecological biochemistry. M.: Mir. 1985. P.176.

[90] Dinan L. Rus. J. // *Plant Physiol*. 1998. V. 45, No. 3. P. 347.

[91] Camps F., Coll J. // *Phytochem*. 1993. V. 32. No. 6. P. 1361.

[92] Dinan L. J. // *Chromatograph*. 1994. V. 658. P. 69.

[93] Inge-Vechtomov S.G. // *Soros educational journal*. 1997. No. 11. P. 16.

[94] Serebriakov E.P., Promonenkov V.K. Methoprene making processes and properties. Results of science and technology. *Series Organic Chemistry*. 1989. V. 9. P.102-167.

[95] Bergamasco R., Horn D.H.S. Invertebrate endocrinology. Vol.1. *Endocrinology of insects* // Eds. R.G.H. Downer, H.N.Y. Laufer, A.R. Liss. New York. 1983. P. 627.

[96] Harris R.L., Iwamura H., Fujita T. // *J. Med. Entomol*. 1984. V. 27. No. 11. P. 1493.

[97] Harris R.L., Frazar E.D., Younger R.L. // *J. Econ. Entomol*. 1974. V. 66. No. 5. P. 1099.

[98] Campbell J.B., Wright J.F. // *J. Econ. Entomol*. 1976. V. 69. No. 5. P. 566.

[99] Younger R.L., Wright J.E., Smalley H.E., Crookshank H.R., Norman J.O. // *J. Med. Entomol.* 1975. V. 12. No. 5. P. 517.

[100] Prasert V., Knapp F. W., Lyons E.T., Drudge J.H. // *J. Econ. Entomol.* 1975. V. 68. No. 5. P. 639.

[101] Mansingh A., Rawlins S.C. // *Naturwissenschaften.* 1977. V.64. No.1. P. 41.

[102] Solomon K.R., Evans A.A. // *J. Med. Entomol.* 1977. V. 14. No. 4. P. 433.

[103] Hopkins D.E., Chamberlain W.F. // *Southwest Entomol.* 1978. V. 3. No. 4. P. 292.

[104] Hong T. K. // *Ann. Appl. Botany.* 1975. V. 80. No. 2. P. 137.

[105] Nickle D.A. // *J. Econ. Entomol.* 1979. V. 72. No. 6. P. 816.

[106] Chippendale G.M., Yin C.M. // *Bull. Entomol.* Res. 1976. V. 66. No.1. P. 75.

[107] Yin C.M., Chaw W.D. // *J. Insect. Physiol.* 1984. V. 30. No. 3. P. 257.

[108] Amine A.M., White G.B. // *Entomol. Exper. Appl.* 1984. V. 36. No.1. P. 69.

[109] Ramakrishnan V. Joshi N.K. // *J. Food Sci Technol.* 1977. V. 14. No.¾. P. 87.

[110] Edwards J.P., Short J.E. // *J. Stored Prod. Res.* 1984. V. 20. No. 1. P. 11.

[111] Mian L.S., Mulla M.S. // *J. Econ. Entomol.* 1982. V. 75. No. 1. P. 80.

[112] Strong R.G., Dickman J. // *J. Econ. Entomol.* 1973. V. 66. No. 5. P. 1167.

[113] Marzke F.O., Coffelt J.A., Silhacek D. L. // *Entomol. Exper. Appl.* 1979. V. 22. No. 3. P. 294.

[114] Loschiavo L.R. // *Manitobo Entomol.* 1975. V. 9. No. 1. P. 43.

[115] Hussein M.H. // *Intern. Pest. Control.* 1983. V. 25. No. 5. P. 140.

[116] Mian L.S., Mulla M.S. // *J. Econ. Entomol. 1983.* V. 76. No. 3. P. 622.

[117] Odinokov V.N., Kukovinets O.S., Zainullin R.A., Tolstikov H.A., Mokrousova Ye.P., Kasradze V.G., Burov V.N. // *Journal Agrochemistry.* 1999. No. 9. P. 60.

[118] Grapov A.F., Pushina M. Ya. // *Journal of D.I.Mendeleev All-Union Chemical Society.* 1988. V. 33. P. 661

[119] Wellinga K., Mulder R., Van Daalen J. J. // *J. Agr. and Food Chem.* 1973. V. 21. P. 348.

[120] Wellinga K., Mulder R., Van Daalen J. J. // *J. Agr. and Food Chem.* 1973. V. 21. P. 993.

[121] Oliver J. E., De Milo A. B., Cohen C. F., Shortino T. J., Robbins W. E. // *J. Agr. and Food Chem.* 1976. V. 24. P. 1065.

[122] Zinoviev V.D., Sokolov M.S., Tischenko Z.A. // *Agrochemistry.* 1994. V. 4. P. 101.

[123] Odinokov V.N., Kukovinets O.S., Lozhkina Ye.A., Bikulova V.Zh., Petrova T.D., Kolesnikova I.V., Platonov V.Ye., Tolstikov H.A. // *News of the USSR Academy of Sciences. Series. Chemistry.* 1992. P.468.

[124] Odinokov V.N., Kukovinets O.S., Lozhkina Ye.A., Tolstikov H.A. From the book "Ecologically and pesticide free technologies for crop production" // Report Abstracts of the All-Russin research-and-production council. Krasnodar. 1994. P.215.

[125] Mulder R., Gijswijt M. // *J. Pestic. Sci.* 1973. V. 4. P. 737.

[126] Nakagawa Y., Sotomatsu T., Irie K., Kitahara K., Iwamura H., Fujita T. // *Pestic. Biochem. Physiol.* 1987. V. 27. N. 2. P. 143.

[127] Sotomatsu T., Nakagawa Y., Fujita T. // *Pestic. Biochem. Physiol.* 1987. V. 27. N 2. P. 156.

[128] Haga T., Toki T., Koyanagi T., Nislriyama R. // *Pestic. Biochem. Physiol.* 1995. V. 10. P. 217.

[129] US Patent 4349681 // *Abstract Journal Chemistry*. 1983. 13O363П.

[130] 60-112770 Japan // *Abstract Journal Chemistry*. 1986. 14O338П.

[131] US Patent 4508722 // *Abstract Journal Chemistry*. 1986. 12O388П.

[132] US Patent 4521426 // *Abstract Journal Chemistry*. 1986. 3O414П.

[133] US Patent 4378991 // *Abstract Journal Chemistry*. 1983. 23O347П.

[134] FRG Patent 3431222 // *Abstract Journal Chemistry*. 1987. 3H152П.

[135] Kitahara K., Nakagawa Y., Nishioka T., Fujita T. // *Agric. Biol. Chem.* 1983. V. 47. No. 7. P. 1583.

[136] FRG Patent 3431222 // *Abstract Journal Chemistry*. 1987. 3H152П.

[137] Yu C. C., Kuhr R. J. // *J. Agr. Food Chem.* 1976. V. 24. No. 1. P. 134.

[138] Hungary Patent 178311 // *Abstract Journal Chemistry*. 1985. T. 6O468П

[139] Great Britain Patent 2118941 // *Abstract Journal Chemistry*. 1984. 14O358П.

[140] US Patent 4452631 // *Abstract Journal Chemistry*. 1985. 6O469П.

[141] Sazonov A.P., Pralial I.I. Biologically active substances // Scientific papers of All-Union Agricultural Academy (VASHKNIL) named after Lenin. M.: *Science*. 1979. P.92.

[142] Benkovskaya G.V., Amirkhanov D.V. // *Agrochemistry*. 1992. V. 4. P. 120.

[143] Japan Patent 6092260 // *Abstract Journal Chemistry*. 1986. 7O335П.

[144] FRG Patent 2601780 // *Chem. Abstr*. 1977. V. 87. 201.

[145] Odinokov V.N., Kukovinets O.S., Lozhkina Ye.A., Tolstikov H.A. // Abstracts of All-Russian conference «Ozone-94». Ufa. 1994. P. 68.

[146] Tolstikov H.A., Odinokov V.N., Kukovinets O.S., Lozhkina Ye.A., Mustafin A.G., Abdrakhmanov I.B. // RF Patent No. 2090556. Reg. 20.09.97 // *Bulletin of inventions*. 1997.No. 27.

[147] Atanassova I. A., Petrov J. S., Mollov N. M. // *Synthesis*. 1987. No.8. P.734.

[148] Fujita T. // *Progr. Phys. Org. Chem.* 1983. V. 14. P. 75.

[149] Atanassova I. A., Petrov J. S., Mollov N. M. // *Synth. Commun.* 1989. V. 19. No. 1. P. 147.

[150] Kozyukov V.P., Feoktistov A.Ye., Mironov V.F. // *Reports Abstracts of the Vth All-Union symposium on organic synthesis*. M. 1988. P. 98.

[151] Swain C. G., Lupton E. C. // *J. Am. Chem. Soc.* 1968. V. 90. No. 16. P. 4238.

[152] Mironov V.F., Kozyukov V.P., Sheludiakov V.D. // *USSR Reports of the Academy of Sciences*. 1978. V. 199. No.1. P.103.

[153] King F.D., Pike S., Walton D. R. M. // *J. Chem. Soc., Chem. Commun.* 1978. No.8. P. 351.

[154] Chang Y. H., Chin F. T., Zon G. // *J. Org. Chem.* 1981. V.46. No.2. P.342.

[155] Maas W., Van Hes R., Grosscurt A.C., Deul D. H. // *Chemie der Pflanzenschutz und Shadligsbekampungsmittel*. Berlin. Heidelberg. New-York. Springer-Verlag. 1981. V. 7. P. 423.

[156] Wie S. I., Sylwester A. P., Wing K. D., Hammock B. D. // *J. Agr. and Food Chem.* 1982. V. 30. No. 5. P. 943.

[157] Lozhkina Ye.A. Ph.D. thesis (Chemistry). Ufa; Institute of Organic Chemistry at the RAN Ural Scientific Center. 1996.

[158] Kukovinets O.S., Lozhkina Ye.A., Odinokov V.N., Tolstikov H.A. // *Bashkir Chemical Journal*. 1996. V.2. No.2. P.78.

[159] Author's certificate. 1016275 USSR // *Bulletin of inventions*. 1983. No. 17.

[160] Odinokov V.N., Ishmuratov G.Yu., Kukovinets O.S., Kharisov R.Ya., Lozhkina Ye.A., Mustafin A.G., Abdrakhmanov I.B., Tolstikov H.A. // *Journal of organic chemistry*. 1998. V. 34. No. 2. P.229.

[161] Odinokov V.N., Serebriakov E.P. *Synthesis of insect pheromones*. Ufa: Gilem. 2001. P.371.

[162] Ishmuratov G.Yu., Kharisov R.Ya., Botsman O.V. et al. // *RAN News. Series. Chemistry*. 2000. No.11. P.1930.

[163] Mikhailyuchenko N.G., Shabalta O.M. / *Technical Cultures*. 1994. No.3-4. P. 7.

[164] Emmanuel Eziashi; Zaets V.G.; Borov L.I. Problems of khapra beetle control / *RUDN. Deposit manuscript.* 1997.01.23 N 15 BC-97. M. P.9.

[165] Applebaum S.W., Gestetner B., Birk Y. // *J. Insect Physiol.* 1965. V. 11. P. 611.

[166] Application 54-101425. 1973. (Japan).

[167] Application 393702. 1989. (Japan).

[168] Melnikov N.N. Pesticides. Chemistry, technology and application. M.: *Chemistry*. 1987. P.149.

[169] Kopezacki P, Gumulka M., Masnyk M., Sarosiek A., BArycki R., Ignacak W. // *Pol. J. Chem.* 2004. V. 78. No. 1. P. 89.

[170] Lavrik P.B., Bartnicki D.E., Feldman J.-G., Sims S.R., Fuchs R.I. // *ACS Symp. Ser.* 1995. V. 695. P. 148.

[171] Dybas R.A., Babu J.R. // Proc. *Brigton Crop Protect. Conf. Pest. Dis.* 1988. V. 2. P. 6.

[172] Diatlova K.A. // *Soros educational journal*. 2001. V. 7. No. 5. P.7.

[173] Grapov A.F. // *Successes of Chemistry*. 1999. V. 68. No.8. P. 773.

[174] Lutova L.A. // *Soros educational journal*. 2000. V.6. No.10. P.10.

[175] Estruch J.J., Carozzi N.B., Desai N., Duck N.B., Warren G.W., Koziel M.G. // *Nature biotechnology*, 1997. V.15. No. 2. P.137.

[176] Valuyeva T.A., Mosolov V.V. // *Successes of Biological Chemistry*. 2002. V. 42. P. 193.

[177] Wilhite S.E., Elden T.C., Brzin J., Smigocki A.C. // *Insect Biochem. Mol. Biol.* 2000. V. 30. No. 12. P. 1181.

[178] Ryan C.A. // *Annu. Rev. Phytopathol.*1990. V. 28. P. 425.

[179] Wolfson J.L., Murdock L.L. // *J. Chem. Ecol.* 1990. V. 16. No. 8. P.1089.

[180] Irie K., Hosoyama H., Takeuchi T., Iwabuchi K., Watanabe H., Abe K., Arai S. // *Plant Mol. Biol.* 1996. V. 30. No. 1. P. 149.

[181] De Boer Jetsko G., Marcel D. // *J. Chem. Ecol.* – 2004. – V. 30. - No. 2. P. 255-271.

[182] Nesar A.K., MdHumaxun K.S. // *Entomon*. 1995. V. 20. No. 3-4. P. 175-182.

[183] Yevsyunin A. // *Chemistry and Life*. 2004. No. 3. P. 40-41.

[184] Baraboi V.A. Vegetable phenols and human health. 1984. M.: *Science*. P.160.

[185] Zakladnoi G.A., Kogteva Ye.F., Vasiliev V.L. // *Plant protection*. 1997. No.10. P. 15.

[186] Dremova V.P., Putintseva L.S., Khodakov P.Ye. Medical disinsection. M.: *Medicine*. 1985. P.226.

[187] Grapov A.F., Pushina M.Ya. // *Journal of D.I.Mendeleev All-Union Chemical Society*. 1988. V. 33. P.661.

[188] Ross R.T. // *Phytoparasitica*. 1996. V. 24. No. 3. P. 246. [*Abstract Journal Biology*. 97.07-04И3.533]

[189] Johnson R.D., Manske D.D. // *J. Pestik Monit*. 1977. V. 11. P. 116.

[190] Polygalina G.V. *Technochemical control of alcoholic beverage industry*. M.: Kolos. 1999. P.327.

Index

A

B

C

D

E

F

G

J

K

L

O

P

Q

R

S

T

U

V